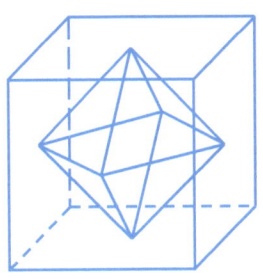 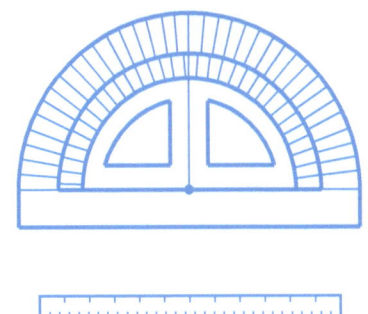 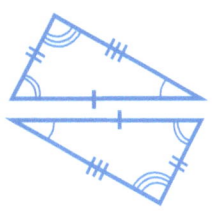 

A Basic Course in Geometry

– Part 4 of 5

Bill Lembke

Citrus Software Publishing

Citrus Ridge, Florida

Published by Citrus Software Publishing, a Division of Citrus Software Corporation

Copyright © 2013 by Citrus Software Corporation

All rights reserved. No part of this publication may be reproduced or distributed in any form or by any means, or stored in a database or retrieval system, without the prior written consent of the publisher, including, but not limited to, network storage or transmission, or broadcast for distance learning.

Printed and bound in the United States of America.

Citrus Software, Citrus Software Publishing, and ABC Method of Instruction are either registered trademarks or trademarks of Citrus Software Corporation in the United Stated and/or other countries. Other products and company names mentioned herein may be the trademarks of their respective owners.

This book expresses the author's views and opinions. The information contained in this book is provided without any express, statutory, or implied warranties. Neither the author, Citrus Software Corporation, nor its resellers, or distributors will be held liable for any damages caused or alleged to be caused either directly or indirectly by this book.

A Basic Course in Geometry – Part 4 of 5

ISBN-13: 978-1477568286

ISBN-10: 147756828X

Table of Contents – Part 4 of 5

Chapter 11 – Spherical Geometry   1

- 11-1 Introduction   1
- 11-2 Terms   1
- 11-3 Properties   3
- 11-4 Spherical Polygons   5
- 11-5 Spherical Angles   6
- 11-6 Spherical Tessellations   7
- 11-7 Spherical Triangle Congruency   8
- 11-8 Spherical Triangle Measurement   9
- 11-9 Summary   10
- Chapter Test   12

Chapter 12 – Geometric Constructions   15

- 12-1 Introduction   15
- 12-2 Why Learn Constructions?   15
- 12-3 Geometry Tools   15
- 12-4 Types of Construction   16
- 12-5 Construction Definitions, Postulates, and Axioms   18
- 12-6 Construction Rules   22
- 12-7 Construction – Traditional Method   23
- 12-8 Construction – Compass Only Method   31
- 12-9 Construction – Modified Traditional Method   32
- 12-10 Summary   35
- Chapter Test   36

Chapter 13 – Geometric Proofs   37

- 13-1 Introduction   37
- 13-2 Why Learn Geometric Proofs?   38
- 13-3 Terms   39
- 13-4 Postulates and Theorems   39

- 13-5 How to Write a Proof … 41
- 13-6 Types of Proofs … 42
- 13-7 Summary … 48
- Chapter Test … 49

Chapter 14 – Assessment … 50

- Chapters 1 – 13 … 50

CHAPTER 11

Spherical Geometry

11-1 Introduction

Spherical geometry is the study of two-dimensional figures on the surface of a sphere. Two-dimensional figures are commonly used in plane geometry and exist on a flat, endless surface. When two-dimensional figures are used in spherical geometry, the surface is curved and not endless; it wraps around on itself. As a result, many of the basic concepts, postulates, and theorems of plane geometry must be modified to work with spherical geometry.

11-2 Terms

Both plane geometry and spherical geometry study two-dimensional objects on a surface. The main difference is the characteristics of the surface. A comparison of geometric terms as use in plane geometry and spherical geometry is made in the table below.

Term	Plane Geometry	Spherical Geometry
1. Point	A point is a location. It does not have a size, width, height, depth, area, or volume. A point can be at any location, such as on a plane or on another object.	A point is a location. It does not have a size, width, height, depth, area, or volume. A point can be at any location, such as on a plane or on another object.
2. Line	A line is a series of points extending infinitely in a straight path in two opposite directions.	A geodesic is a series of points extending in a curved path along a great circle in one direction. A geodesic begins and ends at the same point. Straight lines do not exist.
3. Line Segment	A line segment is part of a line, with points marking the beginning and ending locations.	A geodesic segment is part of a geodesic, with points marking the beginning and ending locations. A geodesic segment is the arc of the great circle between the two points.
4. Ray	A ray is half of a line. It has a point marking the beginning location and extends infinitely in one direction.	A ray is a geodesic. It has a point marking the beginning location and extends along a great circle in one direction

			ending at the beginning point.
5.	Angle	An angle is the space between two rays or line segments that meet at a common end point, called the vertex. An angle can be measured in degrees or radians.	An angle is the space between the planes corresponding to the two great circles that intersect at a common end point, called the vertex. An angle is measured in degrees or radians.
6.	Plane	A plane is a flat surface extending infinitely in all directions. It is made up of points and does not have a thickness.	A plane intersects a sphere through the center to create a great circle.
7.	Surface	The universe is all points on a plane. It extends infinitely in all directions.	The universe is all points on a sphere. It extends a finite distance in all directions.
8.	Plane Intersection	A plane intersecting the surface creates a straight line.	A plane intersecting the surface creates a circle.
9.	Polygon	A polygon is a two-dimensional object bounded by straight lines. The simplest polygon is a triangle with three sides.	A spherical polygon is a two-dimensional object bounded by great circles. The simplest polygon is biangle or lune with two sides.
10.	Triangle Angles	The sum of the interior angles is 180 degrees. A triangle can have one 90 degree angle. The sum of the exterior angles is 360 degrees.	The sum of the interior angles is 180 degrees to 540 degrees. A triangle can have three 90 degree angles. The sum of the exterior angles is 0 degrees to 360 degrees.
11.	Length Measurement	A line segment can be measured in any unit, such as inches or centimeters.	A geodesic segment can only be measured in degrees.
12.	Length	The shortest distance between two points is a straight line.	The shortest distance between two points is a small arc of a great circle. The large arc is not used.
13.	Unique Line	A unique line can be drawn through any two points.	A unique great circle can be drawn through any two points. If the points are antipodal, then an infinite number of great circles can be drawn through the two points.
14.	Two Points	Two points divide a line into one finite section (line segment) and two infinite sections (rays). Two points can be any distance apart, without a greatest	Two points divide a geodesic into two finite sections (geodesic segments). The greatest distance on a sphere is 180 degrees.

		distance.	
15.	Two Lines	Two lines can be parallel or intersect. Two intersecting lines intersect at one point.	Two geodesics can intersect. There are no parallel lines. Two intersecting geodesics (great circles) intersect at two points.
16.	Perpendicular Lines	Two perpendicular lines intersect one time, create four right angles, and create four infinite regions.	Two perpendicular great circles intersect two times, create eight right angles, and create four finite regions.
17.	Three Points	Three points divide a line into two finite sections (line segments) and two infinite sections (rays).	Three points divide a geodesic into three finite sections (geodesic segments). The resulting spherical polygon is a 540 degree spherical triangle.
18.	Three Points on a Line	If A, B, and C are three points on a line, B is between A and C.	If A, B, and C are three points on a great circle, B is between A and C, C is between B and A, and A is between C and B.
19.	Surface Regions	Maximum number of surface regions created with intersecting lines: R = 1 +(N(N + 1))/2	Maximum number of surface regions created with intersecting great circles: R = 2 + (N-1)N
20.	Triangle Points	Lines passing through three points intersect to determine one triangle.	Great circles passing through three points intersect to determine eight spherical triangles.
21.	Similar Triangles	Triangles with congruent corresponding angles are similar, but may not be congruent.	Triangles with congruent corresponding angles are always congruent. Similarity does not exist on a sphere.

Table 11-1: Plane Geometry and Spherical Geometry Term Comparison

• 11-3 Properties

The surface of a sphere is curved, so all lines on a sphere are curved as well. Straight lines can only exist on a flat surface, such as a plane. Because of this, lines on a sphere are called geodesics. A great circle is created by the intersection of a plane and a sphere, with the plane passing through the center of the sphere. It cuts the sphere into two equal parts or hemispheres. The intersection of a plane and a sphere, but not through the center, creates a small circle. The shortest path between two non-antipodal points on the surface of a sphere, known as the

orthodome, lies on a unique great circle that passes through the two points. The arc of the great circle between the two points is called a **geodesic**. If the two points are **antipodal**, then there are an infinite number of shortest paths between them.

Since a sphere is a three dimensional object that can be rotated in any direction, two antipodal points can be designated as **poles** of the sphere to determine the orientation of surface objects. The great circles passing through the poles are known as **lines of longitude** or **meridians**. A great circle perpendicular to the meridians, located halfway between the two poles, is known as the **equator**. Small circles parallel to the equator are known as **lines of latitude**. The length of a great circle is equal to the circumference of the sphere.

The shortest path between two points on the surface of a sphere is an arc of a **great circle**. The two points divide a great circle into two arcs. The shorter distance arc is called the **small arc** or **minor arc** and the longer distance arc is called the **large arc** or **major arc**. The shortest path between two points is the length of the small arc. A small circle is not the shortest path between two points, but is instead a path of constant curvature. There are an infinite number of small circles between any two points. The figures below show antipodal points and orientation of a sphere.

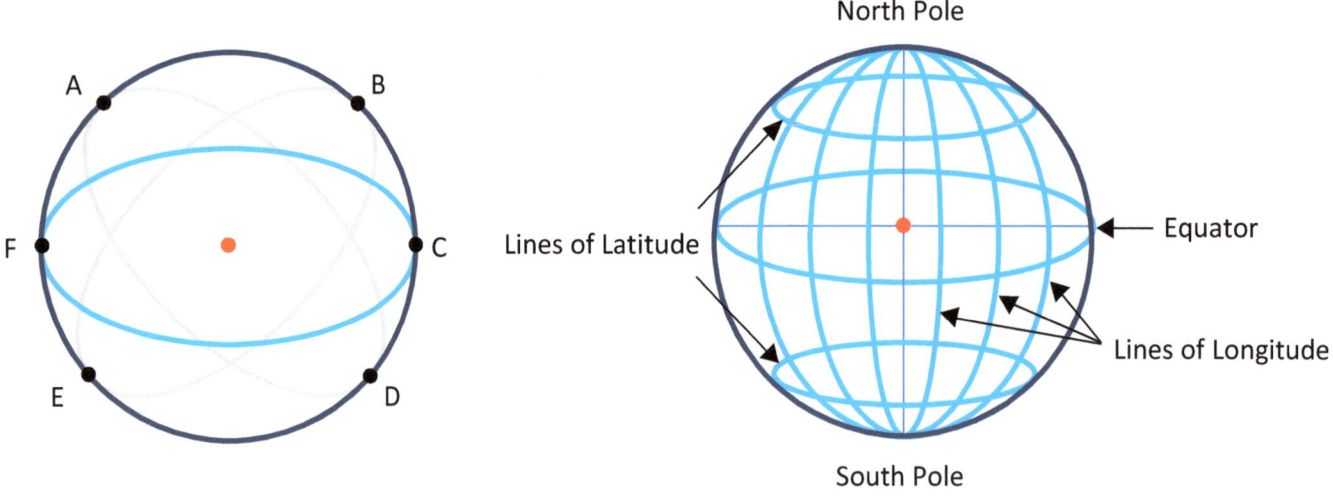

Figure 11-1: Sphere – Antipodal Points   Figure 11-2: Sphere – Object Orientation

Antipodal points are at opposite locations on a great circle. Each great circle has an infinite number of antipodal pairs of points. In the first figure above, the antipodal pairs are A and D, B and E, and C and F. In the second figure above, lines of longitude and latitude are shown. An infinite number of each of these lines can exist. Every line of longitude intersects the equator at a right angle. All lines of latitude, except the equator, are intersected by lines of longitude at angles smaller than a right angle.

11-4 Spherical Polygons

A spherical polygon is a closed two-dimensional object on the surface of a sphere created by the intersection of two or more great circles. The vertices of a spherical polygon are the points of intersection of the great circles and the sides of a spherical polygon are the geodesic segments between the points. The central angle formed at the center of the sphere by the intersection of two great circles that form the sides at a vertex is also called a dihedral angle or spherical angle. The sides are measured by the degrees of their central angle because the sides are arcs of great circles. The angles of a spherical polygon are measured at each vertex between the tangent lines of the two sides forming the vertex angle. A tangent line is extended from the vertex to the maximum curvature of the side to measure the vertex angle because the sides of a spherical polygon are convex.

A spherical biangle is created by the intersection of two great circles. A spherical triangle is created by the intersection of three great circles. A spherical quadrilateral is created by the intersection of four great circles. The plane of each great circle is shown in the figures. The figures below show a spherical biangle, spherical triangle, and spherical quadrilateral.

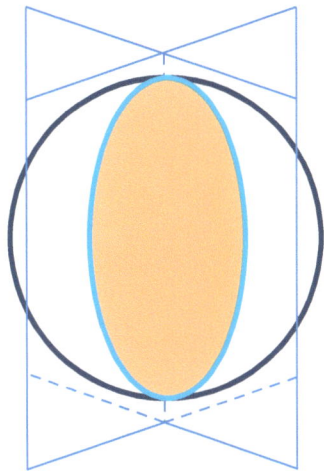 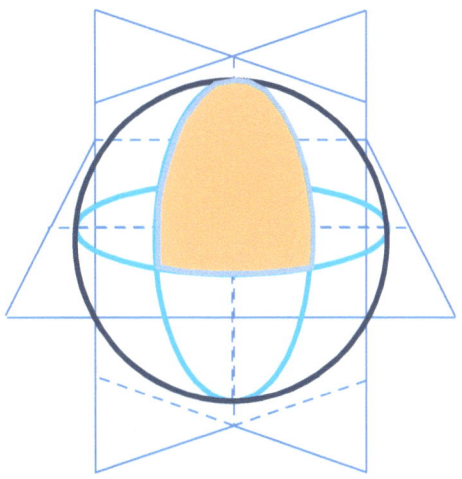 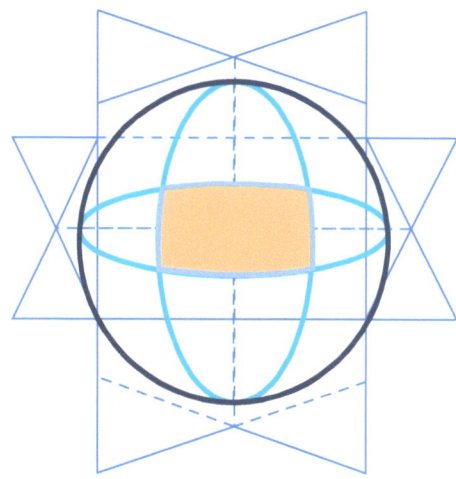

Figure 11-3: Spherical Biangle   Figure 11-4: Spherical Triangle   Figure 11-5: Spherical Quadrilateral

The sum of the interior *n* angles of an *n*-gon on a plane is calculated as: X = (*n*-2)180 degrees. The amount that the sum of the interior *n* angles of an *n*-gon on a sphere exceeds this is called the spherical excess. The spherical excess is the amount by which the sum of the angles of a polygon on a sphere exceeds the sum of the angles of a polygon with the same number of sides in a plane. The spherical excess for a spherical triangle is calculated as: E = a + b + c − 180 degrees or E = a + b + c − π radians, where a, b, and c are the angles of the spherical triangle. The spherical excess is proportional to the spherical area enclosed by the polygon. The figures below compare the surface areas of spherical polygons and plane polygons. As the number of sides of a polygon increases, the amount of spherical excess compared to the corresponding plane polygon decreases.

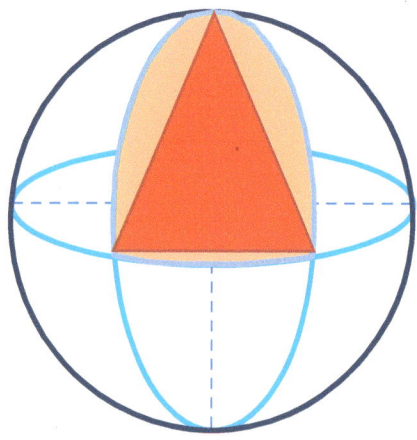

Figure 11-6: Triangle Excess

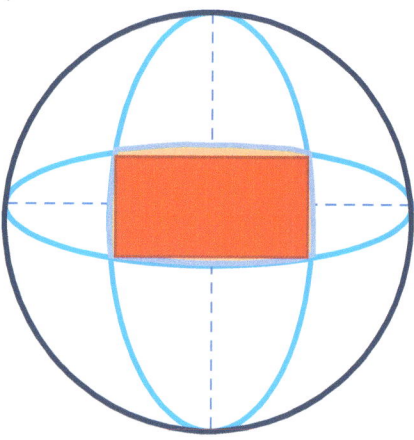

Figure 11-7: Quadrilateral Excess

• 11-5 Spherical Angles

The surface area of a spherical triangle is calculated as: A = π r^2E/180 if the angles are measured in degrees or A = r^2E if the angles are measured in radians, where r is the radius of the sphere. The sum of the angles of a spherical triangle, in degrees, is calculated as: Y = 180(1 + 4 (area of triangle / surface area of sphere)). The area of a spherical polygon with *n* sides is calculated as: A = (y-(*n*-2)180)r^2 or A = (z-(*n*-2)π)r^2, with y the sum of the angles in degrees, z the sum of the angles in radians, and r the radius of the sphere. The total surface area of a sphere is calculated as: S = 4πr^2.

A <mark>regular polygon</mark> has all sides of equal length and all angles of equal measure. Regular polygons can occur on both a plane and a sphere. The sum of the interior angles of a spherical polygon always exceeds the sum of the interior angles of a plane polygon with the same number of sides. As a result, each angle of a spherical polygon will be larger in measure than each angle of a plane polygon with the same number of sides. A comparison of regular polygons angles as use in plane geometry and spherical geometry is made in the table below.

Polygon Name	Number of Sides	Plane Degrees	Spherical Degrees
Biangle	2	0	>0 – 180
Triangle	3	60	>60 – 180
Quadrilateral	4	90	>90 – 180
Pentagon	5	108	>108 – 180
Hexagon	6	120	>120 – 180
Heptagon	7	128.57	>128.57 – 180
Octagon	8	135	>135 – 180
Enneagon	9	140	>140 – 180
Decagon	10	144	>144 – 180

Table 11-2: Plane Geometry and Spherical Geometry Angle Comparison

11-6 Spherical Tessellations

The fraction of the sphere covered by a spherical polygon is calculated as: $F = E/720$ degrees, where the excess is measured in degrees. <mark>Tessellation</mark> is process using shapes to fill an area, like tiling, with no overlapping shapes or gaps. The surface of a sphere can be completely covered with congruent regular spherical polygons. Ignoring biangles and 180 degree angles, there are only five possible regular spherical tessellations. The number of tessellations for both biangles and 180 degree angle is infinite. Spherical polygon tessellations are shown in the table below.

Polygon Name	Spherical Degrees	Polygons at Vertex	Total Polygons
Triangle	72	5	20
Triangle	90	4	8
Triangle	120	3	6
Quadrilateral	120	3	6
Pentagon	120	3	12

Table 11-3: Spherical Polygon Tessellations

The figures below show tessellations of spherical polygons.

   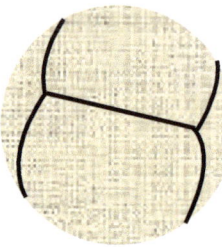

Figure 11-8: 72⁰ Triangles   Figure 11-9: 90⁰ Triangles   Figure 11-10: 120⁰ Triangles   Figure 11-11: 120⁰ Quadrilaterals

 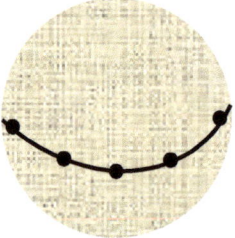 

Figure 11-12: 120⁰ Pentagons   Figure 11-13: Two 180⁰ 12-gons   Figure 11-14: Seven Biangles

• 11-7 Spherical Triangle Congruency

Two triangles are similar if the corresponding angles are congruent or the corresponding sides have lengths that are in the same proportion. Similar triangles have the exact same shape, but may not be the same size. Two triangles are congruent if all pairs of corresponding angles are congruent and all pairs of corresponding sides have the same length. This is a total of six equalities. Congruent triangles have the exact same shape and size. If triangles are congruent, then they are also similar.

The following equality conditions are sufficient to prove congruency for a pair of spherical triangles.

1. (SAS): Two sides in a triangle have the same length as two sides in the other triangle, and the included angles have the same measure.
2. (ASA): Two angles in a triangle have the same measure as two angles in the other triangle, and the included side has the same length.
3. (SSS): Each side of a triangle has the same length as a corresponding side of the other triangle.
4. (AAA): Three angles in a triangle have the same measure as three angles in the other triangle.

On a plane, two triangles with equal angles are similar, but on a sphere the two triangles are congruent.

• 11-8 Spherical Triangle Measurement

Spherical triangles with equal angles must contain the same area because the area is calculated by summing the angles. All spherical triangle with the same measure of angles will contain the same area. The first figure below shows the angles of a spherical triangle. Three mutually perpendicular planes pass through the center of a sphere, cutting the surface of the sphere into eight spherical triangles. The second figure below shows the measurements of a spherical triangle. The vertices of the spherical triangle are points A, B, and C. The central angle, point O, is at the center of the sphere.

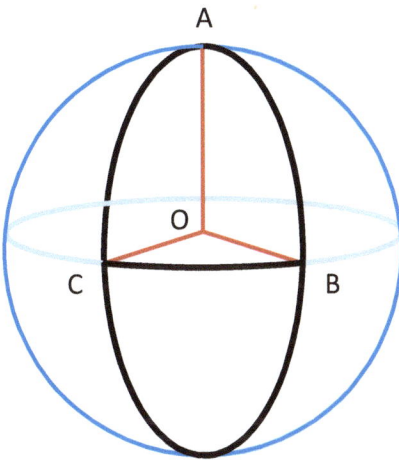

Figure 11-15: Triangle – Angles

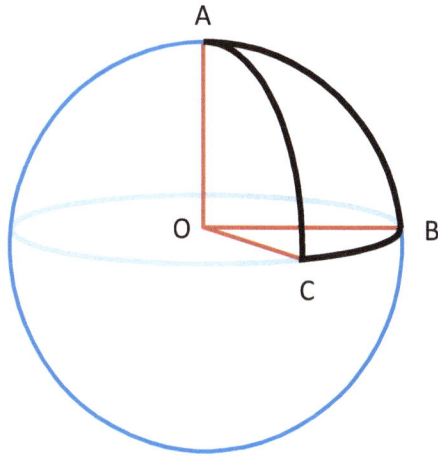

Figure 11-16: Triangle – Measurement

The angles of a spherical triangle are measured at the center of the sphere. This is the location of the plane intersections that create the great circles of the sphere. Angle COB is equal in measure to angle CAB. Angle OCB is equal in measure to angle ACB. Angle OBC is equal in measure to angle ABC. Point A is at the north pole of the sphere. Point B and point C are on the equator of the sphere. Since every line of longitude intersects the equator at a right angle, angle OCB is 90^0 and angle OBC is 90^0. If a value is given for angle COB, such as 45^0, then the perimeter of the spherical triangle can be calculated. The calculation is based on the circumference of a circle, which is: $C = 2\pi r$. A unit circle with a radius of 1 unit can be used.

Arc AB is 90^0 and the length of the arc is calculated as: $90^0/360^0 = (.25)(2)(\pi)(1) = 1.5707$ units. Arc AC is 90^0 and the length of the arc is 1.5707 units. Arc BC is 45^0 and the length of the arc is calculated as: $45^0/360^0 = (.125)(2)(\pi)(1) = 0.7853$ units. The perimeter is the sum of the three side lengths and is calculated as: $1.5707 + 1.5707 + 0.7853 = 3.9267$ units. The circumference of the sphere is calculated as: $(2)(\pi)(1) = 6.2831$ units. The spherical excess for a spherical triangle is calculated as: $E = a + b + c - 180$ or $E = 45 + 90 + 90 - 180 = 45$ degrees. The fraction of the sphere covered by a spherical polygon is calculated as: $F = E/720$ degrees or $45/720 = 0.0625$ or 1/16. The total surface area of a sphere is calculated as: $S = 4\pi r^2$ or 12.5663 square units. The surface of the spherical triangle is calculated as: $12.5663/16 = 0.7853$ square units. The surface area of a spherical triangle, if the angles are measured in degrees, is calculated as: $S = ((A + B + C - 180)/180) \pi r^2 = ((45 + 90 + 90 - 180)/180) \pi r^2 = 0.7853$ square units. The surface area of a spherical triangle, if the angles are measured in radians, is calculated as: $S = (A + B + C - \pi) r^2$. Using the value of the spherical excess, the surface area of a spherical triangle can be calculated as: $S = r^2 E (\pi/180)$ for degrees or $S = r^2 E$ for radians. If a spherical triangle has one vertex at a pole and the other two vertices at the equator, as in this example, and a radius of 1 unit, the length of arc along the equator, BC, will always equal the surface area of the spherical triangle.

11-9 Summary

Spherical geometry is the study of two-dimensional figures on the surface of a sphere. Two-dimensional figures are commonly used in plane geometry and exist on a flat, endless surface. When two-dimensional figures are used in spherical geometry, the surface is curved and not endless; it wraps around on itself. Both plane geometry and spherical geometry study two-dimensional objects on a surface. The main difference is the characteristics of the surface.

The surface of a sphere is curved, so all lines on a sphere are curved as well. Straight lines can only exist on a flat surface, such as a plane. Because of this, lines on a sphere are called geodesics. A great circle is created by the intersection of a plane and a sphere, with the plane passing through the center of the sphere.

Since a sphere is a three dimensional object that can be rotated in any direction, two antipodal points can be designated as poles of the sphere to determine the orientation of surface objects. The shortest path between two points on the surface of a sphere is an arc of a great circle. Antipodal points are at opposite locations on a great circle.

A spherical polygon is a closed two-dimensional object on the surface of a sphere created by the intersection of two or more great circles. The vertices of a spherical polygon are the points of intersection of the great circles and the sides of a spherical polygon are the geodesic segments between the points. A spherical biangle is created by the intersection of two great circles. A spherical triangle is created by the intersection of three great circles. A spherical quadrilateral is created by the intersection of four great circles. The sum of the interior angles of a spherical polygon always exceeds the sum of the interior angles of a plane polygon with the same number of sides.

Tessellation is process using shapes to fill an area, like tiling, with no overlapping shapes or gaps. The surface of a sphere can be completely covered with congruent regular spherical polygons. Ignoring biangles and 180 degree angles, there are only five possible regular spherical tessellations. The number of tessellations for both biangles and 180 degree angle is infinite.

Two triangles are congruent if all pairs of corresponding angles are congruent and all pairs of corresponding sides have the same length. The following equality conditions are sufficient to prove congruency for a pair of spherical triangles: SAS, ASA, SSS, and AAA. On a plane, two triangles with equal angles are similar, but on a sphere the two triangles are congruent.

Spherical triangles with equal angles must contain the same are because the area is calculated by summing the angles. All spherical triangle with the same measure of angles will contain the same area. The angles of a spherical triangle are measured at the center of the sphere. This is the location of the plane intersections that create the great circles of the sphere.

CHAPTER 11 — Chapter Test

Grading Scale: One point for each correct answer.

Excellent = 63-70, Good = 56-62, Average = 49-55, Fair = 42-48, Poor = 0-41

11-2 Terms

Match the term with the definition. Also determine if the definition applies to plane geometry (X), spherical geometry (Y), or both (Z).

A = Point   B = Line   C = Line Segment   D = Ray   E = Angle

F = Plane   G = Surface   H = Plane Intersection   I = Polygon

J = Triangle Angles   K = Length Measurement   L = Length   M = Unique Line

N = Two Points   O = Two Lines   P = Perpendicular Lines

Q = Three Points   R = Three Points on a Line   S = Surface Regions

T = Triangle Points   U = Similar Triangles

1. Lines passing through three points intersect to determine one triangle. ____ ____
2. Lines intersect twice; create eight right angles, and four finite regions. ____ ____
3. Universe is all points, extends a finite distance in all directions. ____ ____
4. A series of points, with points marking beginning and ending locations. ____ ____
5. Two points divide a line into one finite section and two infinite sections. ____ ____
6. If A, B, and C are three points on a line, B is between A and C. ____ ____
7. A location without a size, width, height, depth, area, or volume. ____ ____
8. A two-dimensional object, with simplest type having with two sides. ____ ____
9. A line with a beginning location and extends infinitely in one direction. ____ ____
10. A series of points that begin and end at the same point. ____ ____
11. The shortest distance between two points is a straight line. ____ ____
12. The sum of the interior angles is 180 degrees to 540 degrees. ____ ____

11-3 Properties

Match definitions and terms.

A = Geodesic B = Great Circle C = Hemisphere D = Small Circle
E = Orthodome F = Lines of Longitude or Meridians G = Lines of Latitude
H = Minor Arc I = Major Arc J = Antipodal Points K = Equator

1. Small circles parallel to the equator. _____
2. Intersection of a plane and a sphere, with the plane passing through the center. _____
3. The shortest path between two non-antipodal points on the surface of a sphere. _____
4. Points at opposite locations on a great circle. _____
5. The shorter distance between two points on a great circle. _____
6. Half of a sphere. _____
7. A line on a sphere. _____
8. Great circles passing through the poles. _____
9. The longer distance between two points on a great circle. _____
10. Great circle perpendicular to the meridians, halfway between the two poles. _____
11. Intersection of a plane and a sphere, but not through the center. _____

11-4 Spherical Polygons

Match definitions and terms.

A = Spherical Polygon B = Vertices C = Sides
D = Central Angle E = Spherical Angle F = Spherical Biangle
G = Spherical Triangle H = Spherical Quadrilateral I = Spherical Excess

1. Another name for the central angle. _____
2. Amount the interior angles of an *n*-gon on a sphere exceeds an *n*-gon on a plane. _____
3. A closed two-dimensional object on the surface of a sphere. _____
4. The geodesic segments between the points of intersection of the great circles. _____
5. Polygon created by the intersection of two great circles. _____
6. Formed at the center of the sphere by the intersection of two great circles. _____
7. Polygon created by the intersection of four great circles. _____
8. The points of intersection of the great circles. _____
9. Polygon created by the intersection of three great circles. _____

11-5 Spherical Angles

Calculate the spherical excess, surface area of a spherical triangle, and total surface area of a sphere. The formulas are as follows: $E = a + b + c - 180$. $A = \pi r^2 E/180$. $S = 4\pi r^2$.
(Round to the nearest thousandth.)

1. 72^0 regular triangle, Radius = 5 _____ _____ _____
2. 90^0 regular triangle, Radius = 4 _____ _____ _____
3. 120^0 regular triangle, Radius = 3 _____ _____ _____

11-6 Spherical Tessellations

Calculate fraction of the sphere covered by a spherical polygon. The formula is as follows: $F = E/720$. Write as a decimal and an equivalent fraction. (Round to the nearest thousandth.)

1. 72^0 regular triangle _____ _____
2. 90^0 regular triangle _____ _____
3. 120^0 regular triangle _____ _____
4. 120^0 regular quadrilateral _____ _____
5. 120^0 regular pentagon _____ _____

11-7 Spherical Triangle Congruency

Match definitions and terms.

    A = SSS   B = ASA   C = SSS   D = AAA

1. Three angles in a triangle have the same measure. ____
2. Two angles in a triangle have the same measure and the included side is equal. ____
3. Each side of a triangle is equal to a corresponding side. ____
4. Two sides are equal to corresponding sides and included angles are congruent. ____

11-8 Spherical Triangle Measurement

Calculate the perimeter of a spherical triangle. (Round to the nearest thousandth.)

1. 72^0 regular triangle, Radius = 5 _____
2. 90^0 regular triangle, Radius = 4 _____
3. 120^0 regular triangle, Radius = 3 _____

CHAPTER 12 — Geometric Constructions

12-1 Introduction

Geometric construction is the drawing of lengths, angles, and objects on a plane using drawing tools and rules that govern construction techniques. It is a form of pure geometry studied from an abstract perspective of axioms and postulates along with the properties of objects rather than the measurements of lengths and angles. Each drawing or construction can be proven to be exactly correct based on the rules of construction. The idealized drawing can then be used for practical applications such as architectural designs and manufacturing. There are several types of drawing tools and accompanying rules depending on the construction method chosen. The geometric constructions made with paper and pencil can be simulated in a computer environment using specialized applications such as *The Geometer's Sketchpad* ® and computer aided design (CAD) software. Imagination, creativity, curiosity, and intrepidness can lead to success in problem solving.

This chapter contains 14 constructions problems with step-by-step solutions. To improve understanding, try to solve each problem before checking the provided solution.

12-2 Why Learn Constructions?

Basic knowledge and skills of geometric constructions help students to discover and explore geometric relationships and interpret geometric concepts and theorems. Constructions give the student the opportunity to practice solving problems, which deepens their understanding of geometry. Students become active participants in their learning, rather than passive observers. Constructions give students the ability to solve semi-structured problems and to develop critical thinking skills. It teaches how to think in a clear and logical manner. This prepares students for the reasoning techniques required to develop geometric proofs.

"I hear and I forget. I see and I remember. I do and I understand." – Confucius, Chinese philosopher (551 BCE – 479 BCE)

12-3 Geometry Tools

Greek mathematicians of antiquity began a codification system for geometry. The system took the step from known properties of geometric figures to a system of logic, which could be used to derive unknown properties. Geometric construction was used to create drawings and prove their results. There are four important geometry tools used in geometric construction.

<mark>Compass</mark> – A compass is used to construct circles or arcs. An idealized compass creates perfect circles. A compass established equidistance of points.

<mark>Straightedge</mark> – A straightedge is used to construct straight lines and does not have any distance markings. An idealized straightedge is perfectly straight and of infinite length. A straightedge establishes collinearity of points.

<mark>Ruler</mark> – A ruler is a straightedge with distance markings.

<mark>Protractor</mark> – A protractor is circular or semi-circular ruler used to measure angles.

Geometric constructions are drawn with the simplest possible tools. The figures below show the four geometric tools of construction: compass, straightedge, ruler, and protractor.

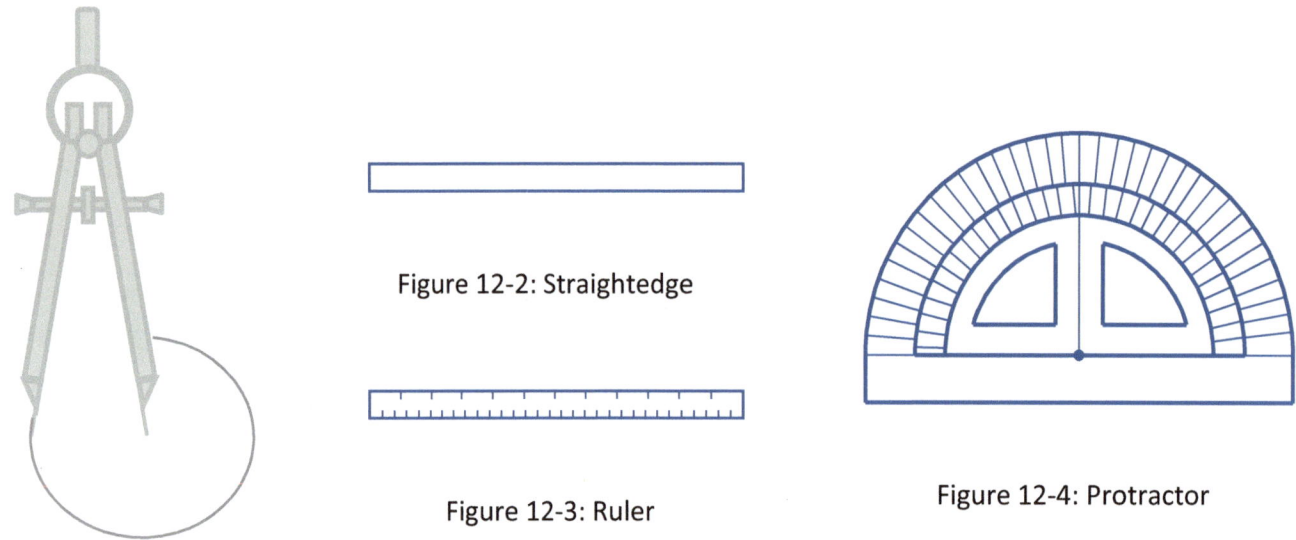

Figure 12-2: Straightedge

Figure 12-3: Ruler

Figure 12-4: Protractor

Figure 12-1: Compass

• 12-4 Types of Construction

There are six types of geometric construction. The traditional, modified traditional, modern traditional, compass only, straightedge only, and modern geometric construction methods are described in the table below.

Name of Construction	Drawing Tools	Description
Traditional or Euclidian	Straightedge, collapsing compass	Straightedge and compass do not have markings. Compass collapses after each use, can not transfer distance.
Modified Traditional	Straightedge, fixed compass	Straightedge and compass do not have markings. Compass can be fixed, can transfer distance.
Modern Traditional	Straightedge, fixed compass	Straightedge and compass have markings. Compass can be fixed, can transfer distance.
Compass Only or Mohr-Mascheroni	Collapsing compass	Compass does not have markings. Compass collapses after each use, can not transfer distance.
Straightedge Only or Poncelet-Steiner	Straightedge	Straightedge does not have markings.
Modern	Straightedge, fixed compass, protractor, ruler, and other tools	All drawing tools have markings. Compass can be fixed, can transfer distance.

Table 12-1: Geometric construction Types

Traditional construction, also known as Euclidean construction, was developed in Greek antiquity in about 300 BCE during the time period of the Greek mathematician Euclid. Mathematicians were influenced by the Greek philosopher Plato's belief that geometry should use no tools but a compass and a straightedge – never measuring instruments such as a marked ruler or a protractor, because they were a workman's tool, not worthy of a scholar. More importantly, measuring instruments were not accurate and their results could not be proven using geometry. The compass and straightedge are idealized versions of the actual tools used. The compass draws perfect circles. The compass is used only to draw circles and arcs through two or more points. The compass does not have any markings and can not be used to transfer distances. The compass is automatically collapsing which means as soon as it was lifted from the drawing surface it closes. The straightedge draws perfectly straight lines. The straightedge does not have any marking and is of infinite length.

Modified traditional construction and modern traditional construction evolved from traditional construction, to allow for easier constructions and the solving of problems not possible with traditional construction. Both of these constructions use a fixed compass which does not collapse after each use. Distances can be transferred by using the compass. Modern traditional construction further allows for measurement markings on the compass and straightedge. Traditional construction can not solve three problems: (1) squaring the circle – drawing a square with the same area as a given circle, (2) doubling the cube – drawing a cube with double the volume of a given cube, and (3) trisecting the angle – dividing a given angle into three smaller angles of the same size. The three problems can be solved using modern traditional construction.

Compass only construction was developed by Danish geometer George Mohr in 1672 and Italian mathematician Lorenzo Mascheroni in 1797. They separately proved that all problems soluble by using of compass and straightedge construction could be solved using a compass alone.

Straightedge only construction was developed by French mathematician Jean Victor Poncelet in 1822 and proved by Swiss mathematician Jacob Steiner in 1833. It was proven that given a fixed circle and its center, all constructions in the plane can be carried out by straightedge alone. It was also proven that the constructions can not be done by straightedge alone. It is impossible to find the center of a circle using a straightedge alone.

Modern construction allows for the use of any geometric tools and all tools have markings. Computer applications can also be used. Very complex and detailed geometric drawings can be made using modern construction techniques. The results are fast and efficient, but may not be provable using geometry.

• 12-5 Construction Definitions, Postulates, and Axioms

Traditional construction began with a strict set of rules based on definitions, postulates, and axioms. A definition is a statement of precise meaning. A postulate is a statement that is accepted as true without proof. An axiom, or common notation, is a self-evident assumption. The following table shows the definitions, postulates, and common notions of Euclid's *Elements*.

Item	Description
Definition 1	A *point* is that which has no part.
Definition 2	A *line* is breathless length.
Definition 3	The extremities of a line are points.
Definition 4	A *straight line* is a line which lies evenly with the points on itself.
Definition 5	A *surface* is that which has length and breadth only.
Definition 6	The extremities of a surface are lines.
Definition 7	A *plane surface* is a surface which lies evenly with the straight lines on itself.
Definition 8	A *plane angle* is the inclination to one another of two lines in a plane which meet one another and do not lie in a straight line.

Definition 9	And when the lines containing the angle are straight, the angle is called *rectilineal*.
Definition 10	When a straight line set up on a straight line makes the adjacent angles equal to one another, each of the equal angles is *right*, and the straight line standing on the other is called *perpendicular* to that on which it stands.
Definition 11	An *obtuse angle* is an angle greater than a right angle.
Definition 12	An *acute angle* is an angle less than a right angle.
Definition 13	A *boundary* is that which is an extremity of anything.
Definition 14	A *figure* is that which is contained by any boundary or boundaries.
Definition 15	A *circle* is a plane figure contained by one line such that all the straight lines falling upon it from one point among those lying within the figure are equal to one another.
Definition 16	And the point is called the *center* of the circle.
Definition 17	A *diameter* of the circle is any straight line drawn through the center and terminated in both directions by the circumference of the circle, and such a straight line also bisects the circle.
Definition 18	A *semicircle* is the figure contained by the diameter and the circumference cut off by it. And the center of the semicircle is the same as that of the circle.
Definition 19	*Rectilineal figures* are those which are contained by straight lines, *trilateral* figures being those contained by three, *quadrilateral* those contained by four, and *multilateral* those contained by more than four straight lines.
Definition 20	Of trilateral figures, an *equilateral triangle* is that which has its three sides equal, an *isosceles triangle* that which has two of its sides alone equal, and a *scalene triangle* that which has its three sides unequal.
Definition 21	Further, of trilateral figures, a *right-angled triangle* is that which has a right angle, an *obtuse-angled triangle* that which has an obtuse angle, and an *acute-angled triangle* that which has its three angles acute.
Definition 22	Of quadrilateral figures, a *square* is that which is both equilateral and right-angled; an *oblong* that which is right-angled but not equilateral; a rhombus that which is equilateral but not right-angled; and a *rhomboid* that which has its opposite sides and angles equal to one another but is neither equilateral nor right-angled. And let quadrilaterals other than these be called *trapezia*.
Definition 23	*Parallel* straight lines are straight lines which, being in the same plane and being produced indefinitely in both directions, do not meet one another in either direction.
Postulate 1	To draw a straight line from any point to any point.
Postulate 2	To produce a finite straight line continuously in a straight line.
Postulate 3	To describe a circle with any center and distance.
Postulate 4	That all right angles are equal to one another.
Postulate 5	That, if a straight line falls on two straight lines make the interior angles on the same side less than two right angle, the two straight lines, if produced indefinitely, meet on that side on which are the angles less than

	the two right angles.
Common Notions 1	Things which are equal to the same thing are also equal to one another.
Common Notions 2	If equals be added to equals, the wholes are equal.
Common Notions 3	If equals be subtracted from equals, the remainders are equal.
Common Notions 4	Things which coincide with one another are equal to one another.
Common Notions 5	The whole is greater than the part.

Table 12-2: Euclid's Definitions, Postulates, and Common Notions

A modern treatment of Euclidean geometry was made by German mathematician David Hilbert in 1899 in *The Foundation of Geometry*. Hilbert's axioms are a set of 21 assumptions, which form a complete set of geometric axioms. They simplify and build upon Euclid's definitions, postulates, and common notions. The 21 assumptions are divided into five groups: incidence axioms, ordering axioms, parallel postulate, congruence axioms, and continuity axioms. The incidence axioms describe collinearity and intersection. Ordering axioms describe the arrangement of points. The parallel postulate describes parallel lines. Congruence axioms describe geometric equivalence. Continuity axioms describe the extension of lines. Hilbert's axioms are shown in the table below.

Item	Description
Incidence 1	For every two points A, B there exists a line a that contains each of the points A, B.
Incidence 2	For every two points A, B there exists no more than one line that contains each of the points A, B.
Incidence 3	There exists at least two points on a line. There exist at least three points that do not lie on a line.
Incidence 4	For any three points A, B, C that do not lie on the same line there exists a plane α that contains each of the points A, B, C. For every plane there exists a point which it contains.
Incidence 5	For any three points A, B, C that do not lie on one and the same line there exists no more than one plane that contains each of the three points A, B, C.
Incidence 6	If two points A, B of a line a lie in a plane α then every point of a lies in the plane α.
Incidence 7	If two planes α, β have a point A in common, then they have at least one more point B in common.
Incidence 8	There exist at least four points which do not lie in a plane.
Order 1	If a point B lies between a point A and a point C then the points A, B, C are three distinct points of a line, and B then also lies between C and A.

Order 2	For two points *A* and *C*, there always exists at least one point *B* on the line *AC* such that *C* lies between *A* and *B*.
Order 3	Of any three points on a line there exists no more than one that lies between the other two.
Order 4	Pasch's Theorem: Any four points *A*, *B*, *C*, *D* of a straight line can always be so arranged that *B* shall lie between *A* and *C* and also between *A* and *D*, and, furthermore, that *C* shall lie between *A* and *D* and also between *B* and *D*. (German mathematician Moritz Pasch in 1882)
Order 5	Pasch's Axiom: Let *A*, *B*, *C* be three points that do not lie on a line and let *a* be a line in the plane *ABC* which does not meet any of the points *A*, *B*, *C*. If the line *a* passes through a point of the segment *AB*, it also passes through a point of the segment *AC*, or through a point of the segment *BC*. (German mathematician Moritz Pasch in 1882)
Congruence 1	If *A*, *B* are two points on a line *a*, and *A'* is a point on the same or on another line *a'* then it is always possible to find a point *B'* on a given side of the line *a'* such that *AB* and *A'B'* are congruent.
Congruence 2	If a segment *A'B'* and a segment *A"B"* are congruent to the same segment *AB*, then segments *A'B'* and *A"B"* are congruent to each other.
Congruence 3	On a line *a*, let *AB* and *BC* be two segments which, except for *B*, have no points in common. Furthermore, on the same or another line *a'*, let *A'B'* and *B'C'* be two segments which, except for *B'*, have no points in common. In that case if *AB*≈*A'B'* and *BC*≈*B'C'*, then *AC*≈*A'C'*.
Congruence 4	If ∠*ABC* is an angle and if *B'C'* is a ray, then there is exactly one ray *B'A'* on each "side" of line *B'C'* such that ∠*A'B'C'*≅∠*ABC*. Furthermore, every angle is congruent to itself.
Congruence 5	Side-Angle-Side Theorem: If for two triangles *ABC* and *A'B'C'* the congruences *AB*≈*A'B'*, *AC*≈*A'C'* and ∠*BAC* ≈ ∠*B'A'C'* are valid, then the congruence ∠*ABC* ≈ ∠*A'B'C'* is also satisfied.
Parallels 1	Let *a* be any line and *A* a point not on it. Then there is at most one line in the plane that contains *a* and *A* that passes through *A* and does not intersect *a*.
Continuity 1	Axiom of Archimedes: If *AB* and *CD* are any segments, then there exists a number *n* such that *n* copies of *CD* constructed contiguously from *A* along the ray *AB* will pass beyond the point *B*. (Greek mathematician Archimedes 273 BCE-212 BCE)
Continuity 2	Line Completeness: To a system of points, straight lines, and planes, it is impossible to add other elements in such a manner that the system thus generalized shall form a new geometry obeying all of the five groups of axioms. In other words, the elements of geometry form a system which is not susceptible of extension, if we regard the five groups of axioms as valid.

Table 12-3: Hilbert's Axioms

Discovery Activity.

1. Discover definitions, postulates, and axioms of construction. Compare Euclid's Definitions, Postulates, and Common Notions with Hilbert's Axioms.
    a. Make a table to show the similarities and differences between the two lists. For example, Euclid's Common Notations 1 is similar to Hilbert's Congruence 2.
    b. Make a table to show which items are not common to both lists.
    c. Make a table to show which items should be added to the lists.
2. Find lists of geometric axioms created by other mathematicians. They are available in print and online from many sources.
    a. Make a table to list of the geometric axioms for each mathematician selected.
    b. Make a table to compare the other geometric axioms to Euclid's and Hilbert's.

Explain why there are differences among the axiom lists. Which list best describes all of the geometric principles and concepts?

12-6 Construction Rules

A compass can be used to draw a circle or arc having a center at a predetermined marked point or an undetermined arbitrary point. A straightedge can be used to draw a straight line, ray, or line segment using one or more predetermined points or an undetermined point. A point can be determined at the intersection of two lines, the intersection of a line and a circle or arc, or the intersection of a circle or arc and a circle or arc. A line segment can be drawn arbitrarily, through or starting at a single point, or through or connecting two points. The radius, arc angle, or line length can not be measured. The construction must be completed using a finite number of steps.

Given one point in a plane, a circle can be drawn, but of unknown size. A line can be drawn through the point, but the slope of the line is unknown and the relationship of the line to other points is unknown. Given two points in a plane, a circle can be drawn with one point as the center and the other point on the boundary. A line can be drawn to connect the two points and the line can be extended infinitely in both directions. While a line can not be drawn arbitrarily tangent to a circle, a line can be drawn through a given point on the boundary of a circle that is tangent to the circle.

All traditional constructions consist of the repeated use of five basic constructions, using the points, lines, and circles that have already been constructed. The five basic constructions are as follows: (1) Construct a line through two given points, (2) Construct a circle with a given center and given radius, (3) Determine the point of intersection of two given lines, (4) Determine the points of intersection of a given line and a given circle, and (5) Determine the points of intersection of two given circles. The five basic constructions are shown in the figures below.

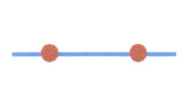

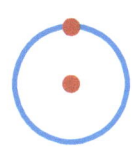

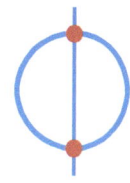

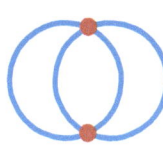

Figure 12-5: Basic Construction 1

Figure 12-6: Basic Construction 2

Figure 12-7: Basic Construction 3

Figure 12-8: Basic Construction 4

Figure 12-9: Basic Construction 5

- **12-7 Construction – Traditional Method**

Problem 1 – Bisect an Angle.

1. Given: Angle BAC.
2. Draw circle with center A.
3. Draw circle with center D, radius A.
4. Draw circle with center E, radius A.
5. Mark F at intersection of circles D and E.
6. Draw line AF.
7. Angle BAC is bisected. BAF ≅ FAC.

Table 12-4: Problem 1

Figure 12-10: Problem 1

Problem 2 – Draw a Perpendicular Line.

1. Given: Line DC.
2. Draw circle with center A.
3. Draw circle with center B, radius A.
4. Mark E and F at intersection of circles A and B.
5. Draw line EF.
6. Line EF is perpendicular to line DC at G.

Table 12-5: Problem 2

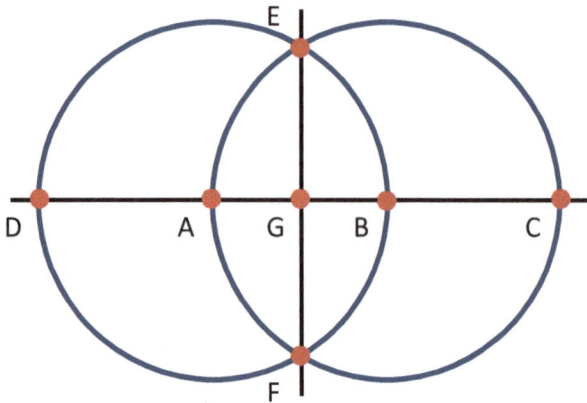

Figure 12-11: Problem 2

Problem 3 – Draw an Equilateral Triangle.

1. Given: Line Segment AB.
2. Draw circle with center A, radius B.
3. Draw circle with center B, radius A.
4. Mark C at intersection of circles A and B.
5. Draw line segments AC and BC.
6. Triangle ABC is an equilateral Triangle.

Table 12-6: Problem 3

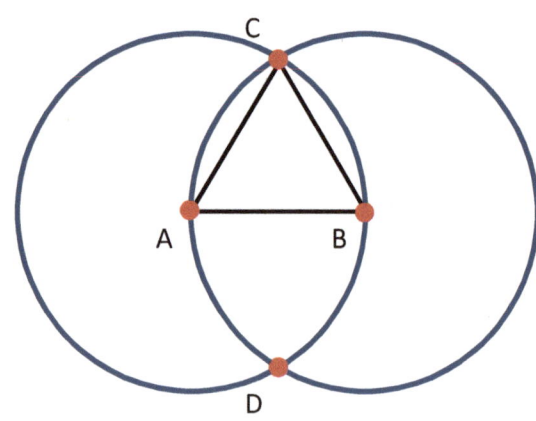

Figure 12-12: Problem 3

Problem 4 – Copy a Line Segment at a Given Point.

1. Given line segment AB and point C.
2. Draw line segment BC.
3. Draw equilateral triangle on BC.
4. Extend CD and DB.
5. Draw circle with center B, radius A.
6. Draw circle with center D, radius F.
7. BA = BF, DF = DG.
8. Since DB = DC, then BF = CG.
9. Therefore, BA = CG.
10. BA has been copied to CG.

Table 12-7: Problem 4

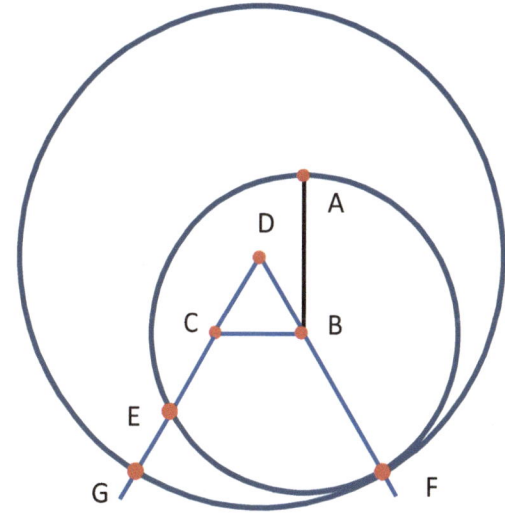

Figure 12-13: Problem 4

Problem 5 – Trisect a Given Arbitrary Angle.

Note: This is one of the three geometric problems from Greek antiquity which was deemed to be impossible using traditional construction methods. Since about 300 BCE, mathematicians have been unable to solve this problem. The solution to the problem was algebraically proved impossible by French mathematician Pierre Wantzel (1836). The following solution created by American mathematician Bill Lembke (2011) shows this problem can be solved. [This solution has been reviewed by mathematicians and has been determined to be a valid solution.]

1. Given: Angle BAC.
2. Draw circle with center A.
3. Mark B and C.
4. Draw line BC.

Table 12-8: Problem 5A

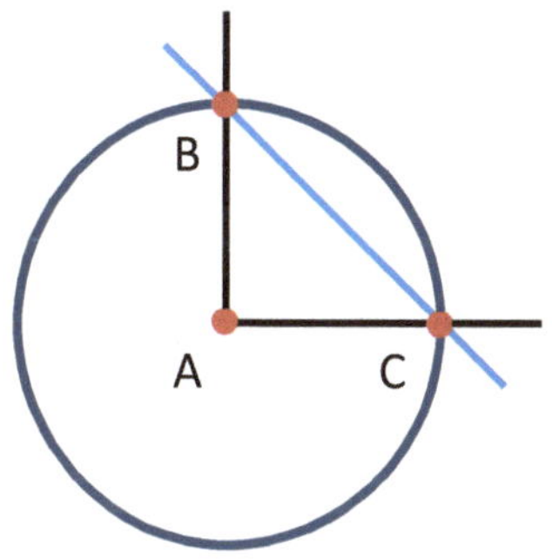

Figure 12-14: Problem 5A

25

5. Draw circle with center B.
6. Draw circle with center D, radius B.
7. Draw circle with center E, radius B.
8. Draw circle with center D, radius E.
9. Draw circle with center E, radius D.
10. Mark F and G.
11. Draw line FG.

Table 12-9 Problem 5B

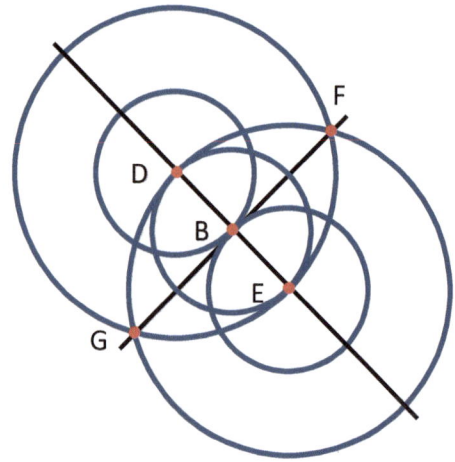

Figure 12-15: Problem 5B

12. Draw circle with center C.
13. Draw circle with center H, radius C.
14. Draw circle with center I, radius C.
15. Draw circle with center H, radius I.
16. Draw circle with center I, radius H.
17. Mark J and K.
18. Draw line JK.

Table 12-10: Problem 5C

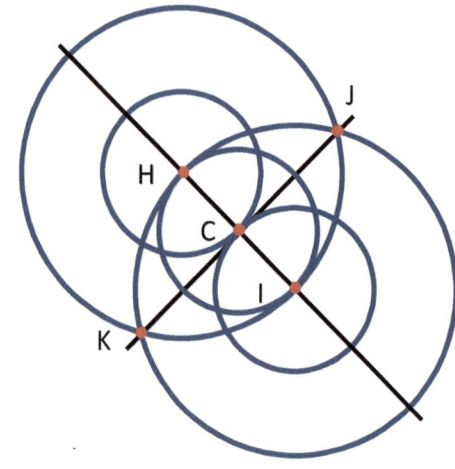

Figure 12-16: Problem 5C

19. Draw circle with center B.
20. Draw circle with center L, radius B.
21. Draw circle with center M, radius B.
22. Mark N at intersection of circles L and M.
23. Draw line BN.
24. Angle LBM is bisected. LBN ≅ NBM.

Table 12-11: Problem 5D

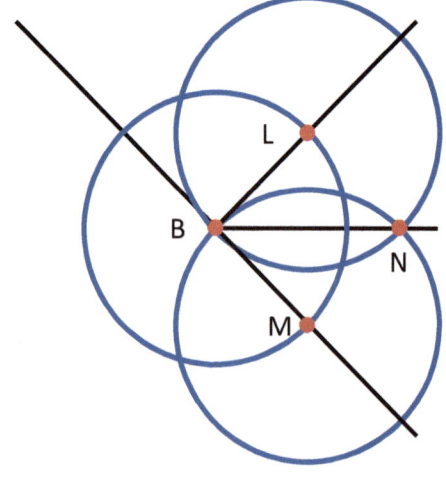

Figure 12-17: Problem 5D

25. Draw circle with center C.
26. Draw circle with center O, radius C.
27. Draw circle with center P, radius C.
28. Mark Q at intersection of circles O and P.
29. Draw line CQ.
30. Angle OCP is bisected. OCQ ≅ QCP.

Table 12-12: Problem 5E

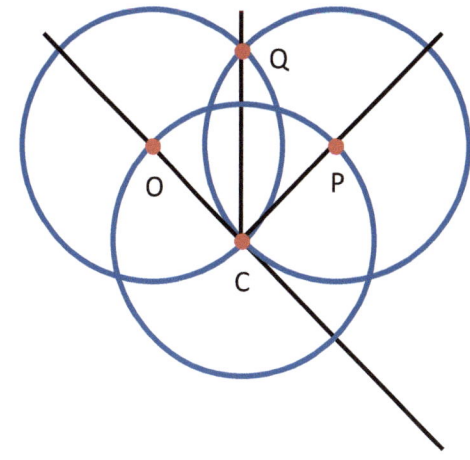

Figure 12-18: Problem 5E

31. Mark R at intersection of lines BN and CQ.
32. Angle BRC is a right angle.
33. Erase circles.

Table 12-13: Problem 5F

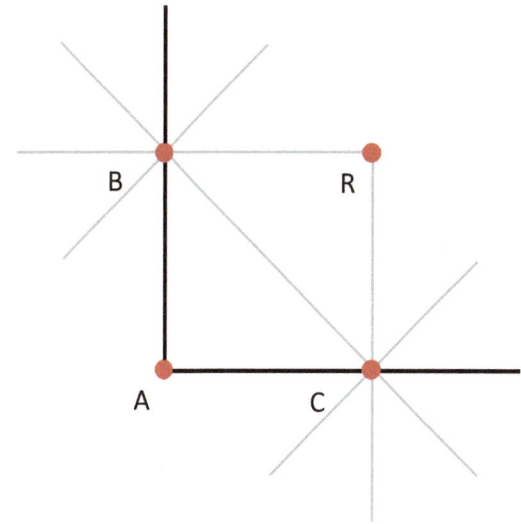

Figure 12-19: Problem 5F

34. Mark S on line RC.
35. Draw circle with center R, radius S.
36. Draw circle with center S, radius R.
37. Mark T at intersection of circles R and S.
38. Draw line segments RT and ST.
39. Triangle RST is an equilateral Triangle.
40. Angle BRT is 30 degrees.
41. Angle TRS is 60 degrees.

Table 12-14: Problem 5G

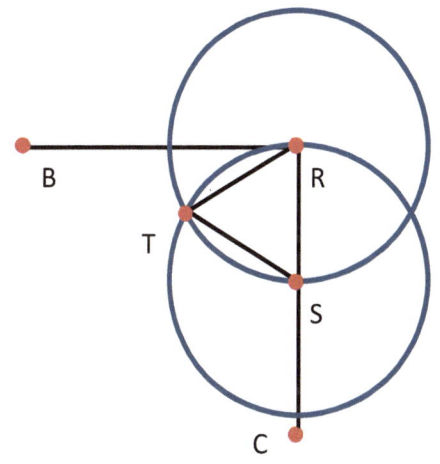

Figure 12-20: Problem 5G

42. Erase circles.
43. Draw circle with center R, radius S.
44. Draw circle with center S, radius R.
45. Draw circle with center T, radius R.
46. Mark U at intersection of circles S and T.
47. Draw line RU.
48. Angle TRC is bisected. TRU ≅ URS.
49. Angle BRC is trisected.
50. Angles BRT, TRU, and URC are 30 degrees.

Table 12-15: Problem 5H

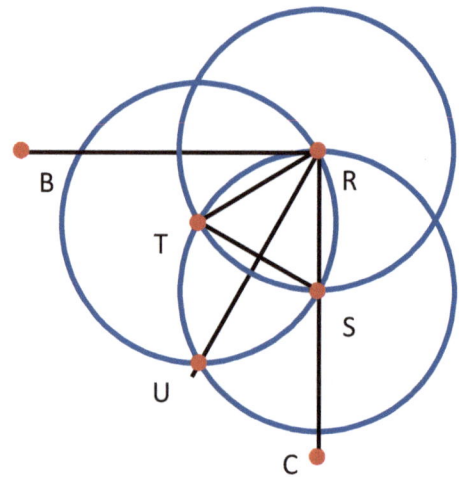

Figure 12-21: Problem 5H

51. Erase circles. Erase ST.
52. Extend RT to intersect BC at V.
53. Extend TU to intersect BC at W.
54. Draw line segments VA and WA.
55. Angle BAC is trisected.
56. Angles BAV, VAW, and WAC are 30 degrees.
57. Angle BAC is 90 degrees. Problem is solved.

Table 12-16: Problem 5I

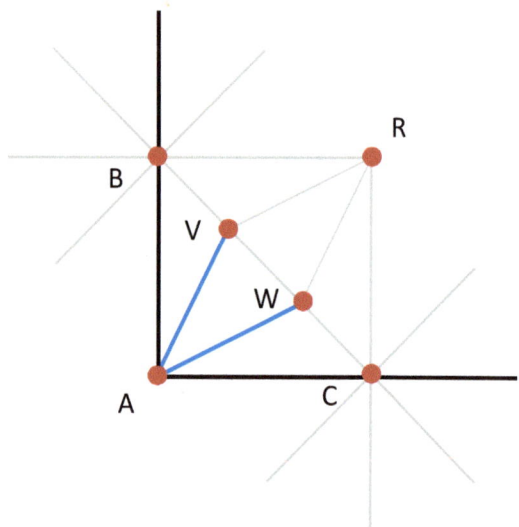

Figure 12-22: Problem 5I

58. Proof: The measure of the angle can increase and decrease as the vertex moves along the angle bisector line of angle BAC.
59. As the vertex location changes, line AB will pivot at B and line AC will pivot at C.
60. The arbitrary angle BAC remains trisected by line segments VA and WA.

Table 12-17: Problem 5J

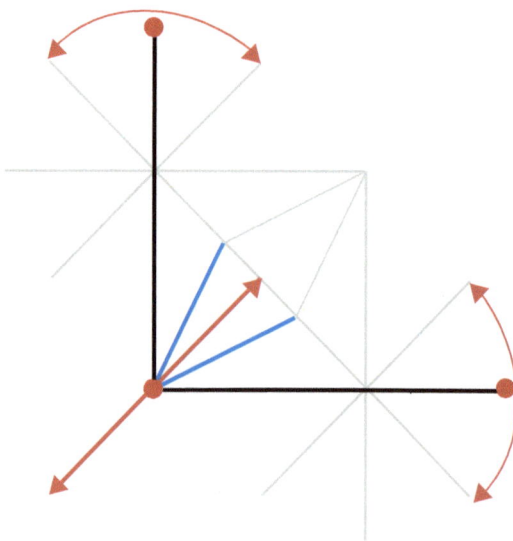

Figure 12-23: Problem 5J

Problem 6 – Find the Center of a Circle.

1. Given: Circle.
2. Draw line AB across circle.
3. Mark A and B at intersections of circle and line.
4. Draw circle with center A, radius B.
5. Draw circle with center B, radius A.
6. Mark C and D at intersections of circles A and B.
7. Draw line CD.
8. Mark E at intersection of CD and bottom of original circle.
9. CD is a perpendicular bisector of AB.

Table 12-18: Problem 6A

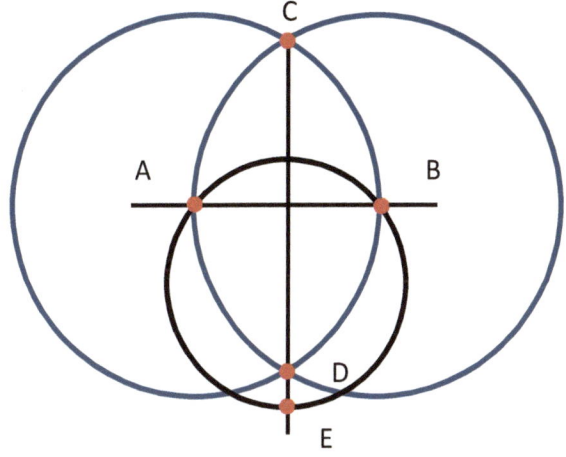

Figure 12-24: Problem 6A

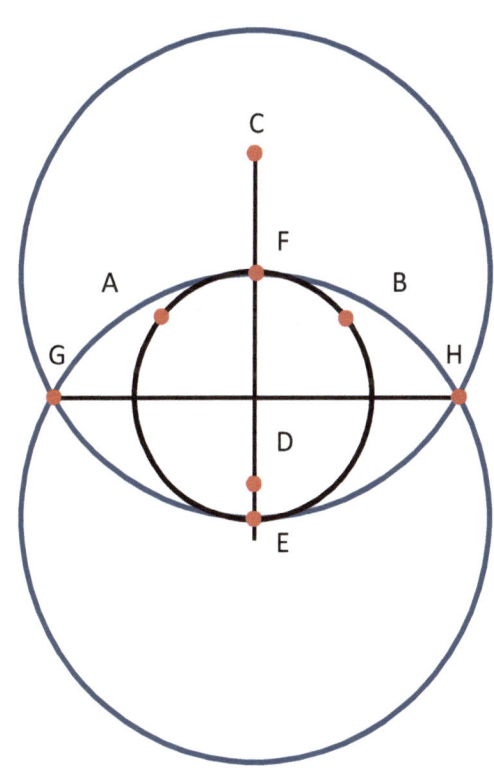

10. Erase circles A and B. Erase AB.
11. Mark F at intersection of CD and top of original circle.
12. Draw circle with center E, radius F.
13. Draw circle with center F, radius E.
14. Mark G and H at intersections of circles E and F.
15. Draw line segment GH.

Table 12-19: Problem 6B

Figure 12-25: Problem 6B

16. Erase circles E and F.
17. Mark I at intersection of CD and GH.
18. I is the center of the original circle.

Table 12-20: Problem 6C

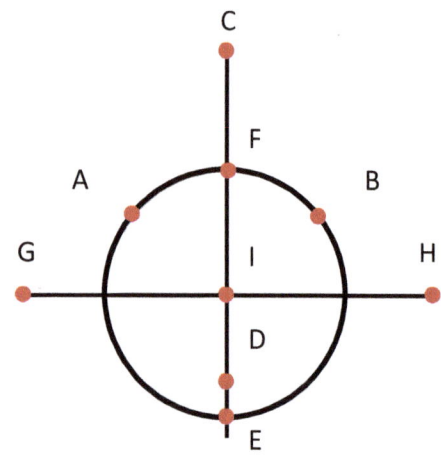

Figure 12-26: Problem 6C

19. Steps 1-18.
20. Complete drawing shown.

Table 12-21: Problem 6D

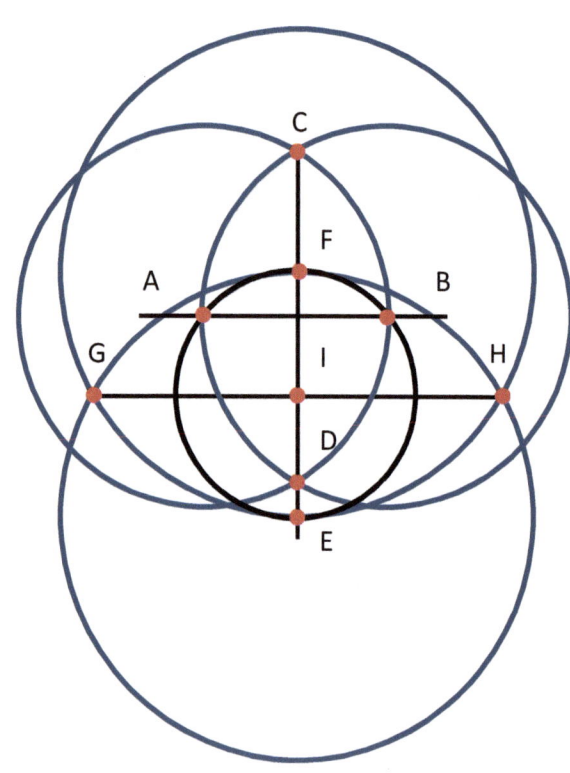

Figure 12-27: Problem 6D

Problem 7 – Draw a Parallel Line.

1. Given: Line segment AB and point C.
2. Draw line BC.
3. Draw circle with center B, radius C.
4. Mark D at intersection of circle and BC.
5. Draw line AD.
6. Draw circle with center A, radius D.
7. Mark E at intersection of circle A and AD.
8. Draw CE.
9. AB and CE are parallel lines.

Table 12-22: Problem 7

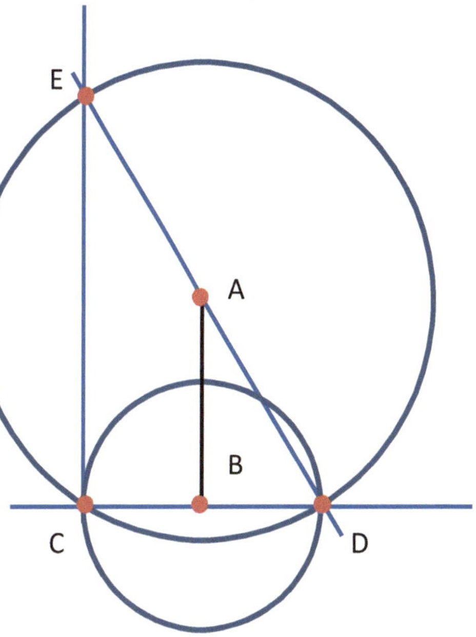

Figure 12-28: Problem 7

12-8 Construction – Compass Only Method

Problem 8 – Copy a Circle.

1. Given: Circle A.
2. Draw circle A2 with center A, any radius greater than circle A.
3. Mark B at any location on circle A2.
4. Draw circle with center B, radius A.
5. Mark C at intersection of circles A2 and B.
6. Mark D at intersection of circles A and B.
7. Draw circle with center C, radius D.
8. Mark E at intersection of circles A2 and C.
9. Draw circle B2 with center B, radius E.
10. Circles A and B2 have the same radius.

Table 12-23: Problem 8

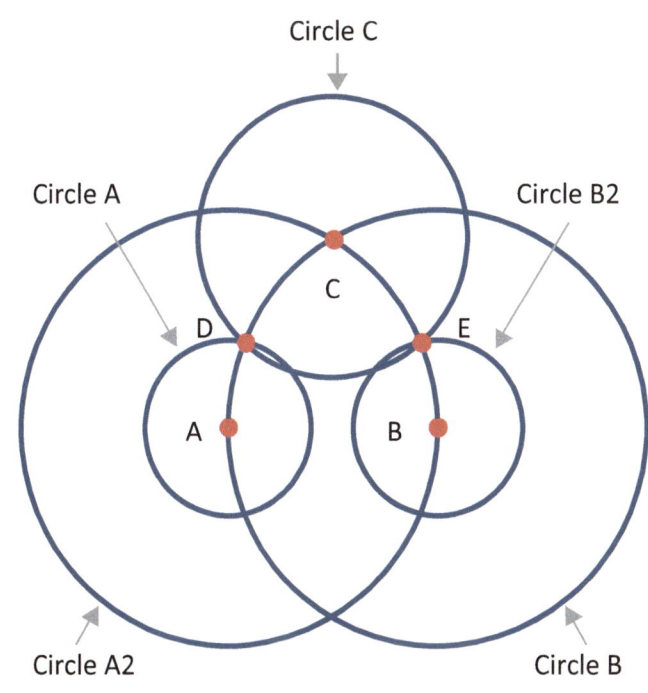

Figure 12-29: Problem 8

12-9 Construction – Modified Traditional Method

Problem 9 – Copy a Line Segment at a Given Point.

1. Given line segment AB and point C.
2. Set compass point at A, adjust width to B.
3. Without changing compass width, set compass point at C and draw circle.
4. Mark D at any location on circle.
5. Draw line segment CD.
6. BA has been copied to CD.

Table 12-24: Problem 9

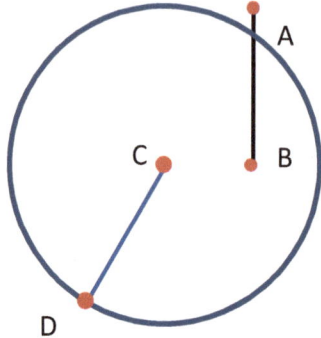

Figure 12-30: Problem 9

Problem 10 – Draw a Parallel Line.

1. Given: Line segment AB and point C.
2. Draw line from C across AB.
3. Mark D at intersection of line and AB.
4. Draw circle with center C, radius less than half of C to D distance.
5. Draw circle with center D, same radius.
6. Mark E and F at intersections of circle C and line CD.
7. Mark G and H at intersections of circle D and line CD.
8. Mark I and J at intersections of circle D and AB.
9. With compass, measure the distance from G to J.
10. Draw circle with center F, radius GJ.
11. Mark K and L at intersections of circle F and circle C.
12. Draw line CK.
13. AB and CK are parallel lines.

Table 12-25: Problem 10

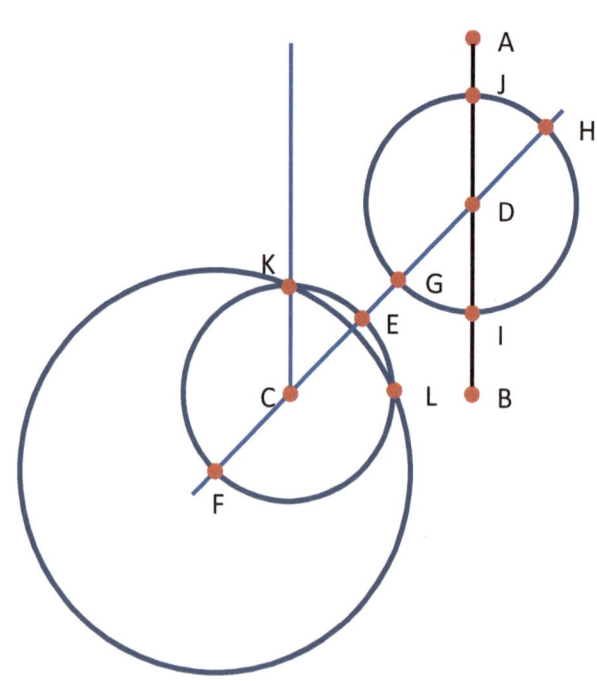

Figure 12-31: Problem 10

Problem 11 – Copy a Circle.

1. Given: Circle A.
2. With compass, measure the radius of circle A.
3. Mark B at arbitrary location.
4. Draw circle with center B, radius A.
5. Circles A and B have the same radius.

Table 12-26: Problem 11

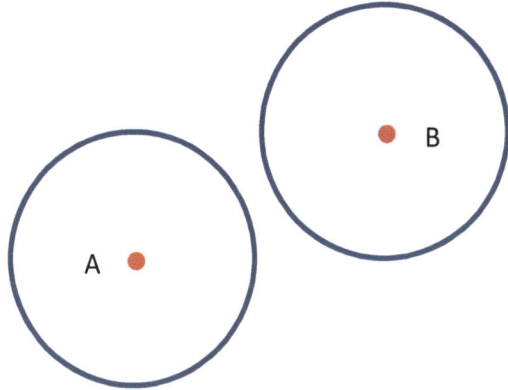

Figure 12-32: Problem 11

Problem 12 – Copy an Angle.

1. Given: Angle BAC.
2. Mark D at arbitrary location.
3. From D draw ray DE.
4. Draw circle with center A.
5. Mark F and G at intersection of circle and angle sides.
6. Draw circle with center D, radius A.
7. Mark H at intersection of circle D and DE.
8. With compass, measure the distance from F to G.
9. Draw circle with center H, radius FG.
10. Mark I and J at intersection of circles D and H.
11. From D draw ray DI.
12. Angle BAC and IDE are congruent.

Table 12-27: Problem 12

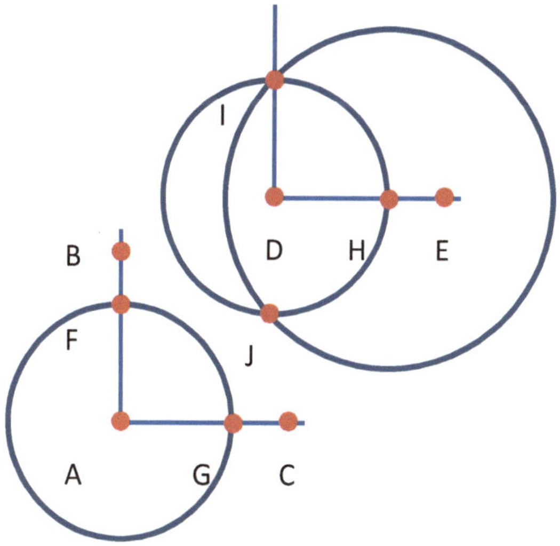

Figure 12-33: Problem 12

Problem 13 – Copy a Triangle.

1. Given: Triangle ABC.
2. Mark D at an arbitrary location.
3. With compass, measure the distance from B to C.
4. Draw circle with center D, radius BC.
5. Mark E at any location on circle.
6. With compass, measure the distance from B to A.
7. Draw circle with center D2, radius BA.
8. With compass, measure the distance from A to C.
9. Draw circle with center E, radius AC.
10. Mark F at intersection of circles D2 and E.
11. Draw line segments DE, EF, and FD.
12. Triangles ABC and DEF are congruent.

Table 12-28: Problem 13

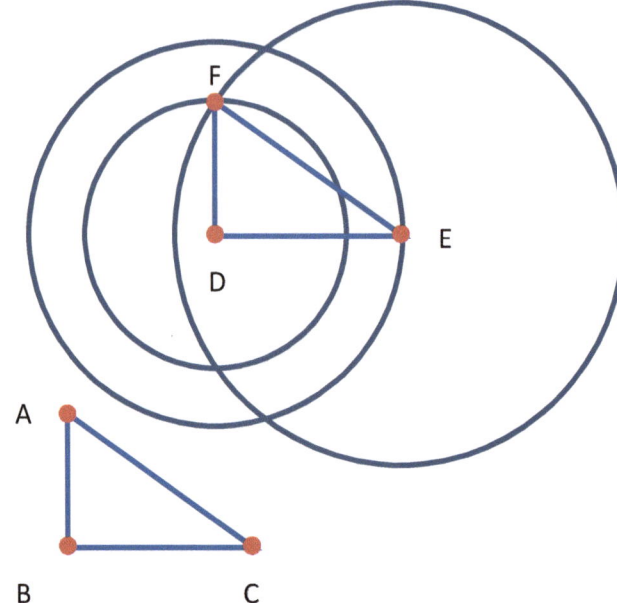

Figure 12-34: Problem 13

Problem 14 – Draw a Hexagon, Given One Side.

1. Given: Line segment AB.
2. Draw circle with center A, radius B.
3. Draw circle with center B, radius A.
4. Mark C and D at intersections of circles.
5. Draw circle with center C, radius A.
6. Mark E and F at intersections of circles A and B with circle C.
7. Draw circle with center E, radius A.
8. Mark G at intersection of circles E and C.
9. Draw circle with center G, radius A.
10. Mark H at intersection of circles G and C.
11. Draw circle with center H, radius A.
12. Draw circle with center F, radius A.
13. Draw line segments AE, EG, GH, HF, and FB.
14. C is the circumcenter of the hexagon.

Table 12-29: Problem 14

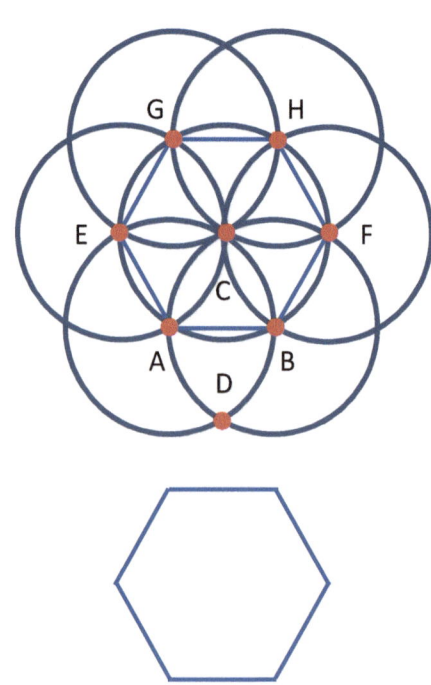

Figure 12-35: Problem 14

12-10 Summary

Geometric construction is the drawing of lengths, angles, and objects on a plane using drawing tools and rules that govern construction techniques. It is a form of pure geometry studied from an abstract perspective of axioms and postulates along with the properties of objects rather than the measurements of lengths and angles. Each drawing or construction can be proven to be exactly correct based on the rules of construction.

Basic knowledge and skills of geometric constructions help students to discover and explore geometric relationships and interpret geometric concepts and theorems. Constructions give the student the opportunity to practice solving problems, which deepens their understanding of geometry. Greek mathematicians of antiquity began a codification system for geometry. The system took the step from known properties of geometric figures to a system of logic, which could be used to derive unknown properties. Geometric construction was used to create drawings and prove their results.

Traditional construction, also known as Euclidean construction, was developed in Greek antiquity in about 300 BCE during the time period of the Greek mathematician Euclid. Modified traditional construction and modern traditional construction evolved from traditional construction, to allow for easier constructions and the solving of problems not possible with traditional construction.

Traditional construction began with a strict set of rules based on definitions, postulates, and axioms. A compass can be used to draw a circle or arc having a center at a predetermined marked point or an undetermined arbitrary point. A straightedge can be used to draw a straight line, ray, or line segment using one or more predetermined points or an undetermined point.

All traditional constructions consist of the repeated use of five basic constructions, using the points, lines, and circles that have already been constructed. The five basic constructions are as follows: (1) Construct a line through two given points, (2) Construct a circle with a given center and given radius, (3) Determine the point of intersection of two given lines, (4) Determine the points of intersection of a given line and a given circle, and (5) Determine the points of intersection of two given circles.

CHAPTER 12

Chapter Test

Grading Scale: One point for each correct answer.

Excellent = 8-9, Good = 7, Average = 6, Fair = 5, Poor = 0-4

• 12-3 Geometry Tools

Match definitions and terms.

    A = Compass   B = Straightedge   C = Ruler   D = Protractor

1. Used to construct straight lines and has distance markings.   ____
2. Used to construct straight lines and does not have any distance markings.   ____
3. Used to measure angles.   ____
4. Used to construct circles or arcs.   ____

• 12-4 Types of Construction

Match definitions and terms.

    A = Traditional   B = Modified Traditional   C = Modern Traditional
    D = Compass Only   E = Modern

1. Straightedge and fixed compass have markings, can transfer distance.   ____
2. Straightedge and collapsing compass do not have markings, can not transfer.   ____
3. Straightedge, fixed compass, protractor, ruler, and other tools.   ____
4. Straightedge and fixed compass do not have markings, can transfer distance.   ____
5. Collapsing compass does not have markings, can not transfer.   ____

CHAPTER 13

Geometric Proofs

13-1 Introduction

A ==geometric proof== is a series of statements based on definitions, postulates, axioms, and theorems used to show that a mathematical statement is valid or true. An unproven proposition or hypothesis that is believed to be true is known as a ==conjecture==. Once a conjecture is proven to be true, it becomes a ==theorem==. Proofs use deductive reasoning or logic to evaluate deductive arguments. An ==argument== is collection of reasons or evidence used to prove a conjecture. An example of a deductive argument is as follows:

1. Polygons have straight sides. (statement)
2. Triangles are polygons. (statement)
3. Therefore, triangles have straight sides. (conclusion)

The first statement declares objects classified as polygons have an attribute of straight sides. The second statement declares triangles to be a member of the group of objects classified as polygons. The third statement concludes that triangles have the polygon attribute of straight sides.

Two forms of deductive reasoning are the use of detachment and syllogism. A conditional statement contains a hypothesis and a conclusion. ==Detachment== uses one conditional statement and a statement to reach a conclusion. An example of a deductive argument using detachment is as follows:

1. If a square has four sides, then it is a quadrilateral. (conditional statement)
2. A square has four sides. (statement)
3. A square is a quadrilateral. (conclusion)

==Syllogism== uses two conditional statements and reaches a conclusion by combining the hypothesis of one statement with the conclusion of another statement. An example of deductive argument using syllogism is as follows:

1. If a polygon has three sides, then it is a triangle. (conditional statement)
2. If a polygon is a triangle, then it has three angles. (conditional statement)
3. If a polygon has three sides, then it has three angles. (conclusion)

If the statements of a deductive argument are not valid, then the conclusion will be invalid or false. An example of a deductive argument based on invalid statements is as follows:

1. All polygons have six sides. (statement)
2. A triangle is a polygon. (statement)

3. Therefore, a triangle has six sides.   (conclusion)

The conclusion of the above deductive argument can be proven to be false. The validity of the statements will have an effect on the soundness of the conclusion. The conclusion must be true for all cases for the conjecture to be proven valid.

This chapter contains 7 geometric proof problems with step-by-step solutions. To improve understanding, try to solve each problem before checking the provided solution.

13-2 Why Learn Geometric Proofs?

Geometric proofs are often considered the ultimate goal of learning geometry. Proofs formally present the results of mathematical thought using logic and deductive reasoning as a method for establishing the validity of ideas. Without a solid foundation in geometry, students may be unprepared to successfully develop proofs. The most effective path to developing meaningful proofs is to first guide students to learn significant and interesting geometric concepts. This will allow students to use visual justification and empirical thinking as a foundation for higher geometric thought. Students begin to develop and test their own ideas based on perceived connections and associations among learned concepts. As students are required to explain and justify their ideas, this will encourage the students to refine their thinking and will gradually lead them to understand the need for formal proofs. As students begin to develop their own proofs, the instruction of formal proofs becomes most effective. For this reason, formal proofs should be taught at the end of a geometry course.

Reasoning and proof offer students effective ways of communicating mathematical ideas and insights. These components, especially the ability to reason, are integral parts of understanding mathematics. They help the student to note and express patterns, structure, and similarities in mathematics and real-world situations as well. Students will learn to that reasoning and proof are a fundamental aspect of geometry. This will allow the students to make and investigate conjectures, develop and evaluate arguments and proofs, and select and use various types of reasoning and methods of proof.

Learning geometric proofs will develop analytical thinking skills. Constructing geometric proofs requires complex organization skills to manage given information, create drawings, and to ensure the progress throughout the task. Proofs require logical thinking and a higher level of thinking to gather and use all of the relevant geometric definitions, postulates, axioms, and theorems in a valid deductive argument that proves a mathematical statement. The value of education is not just in the memorization of facts and figures, but is how the knowledge is used as a tool in a creative process to develop new ideas. Analytical thinking uses an ordered process and reasoning based on evidence and logic. Logical reasoning is the basis for geometric proofs. The skills needed to develop geometric proofs will enable the student to be successful in other areas of mathematics and subjects.

13-3 Terms

A part of the process in creating geometric proofs requires terms to be defined. The following list of terms is important for geometric proofs.

Auxiliary Lines – Lines that are added to a given drawing which help to demonstrate something and prove a statement of a proof.

Contradiction – A statement that is always false. If x=1 is a true statement, then the statement x≠1 is a contradiction or false statement.

Direct Proof – A type of proof in which the conclusion is shown to be true directly from statements of the proof. Direct proofs are the most common type of proof.

Geometric Poof – A series of statements based on definitions, postulates, axioms, and theorems used to show that a mathematical statement is valid or true.

Indirect Proof – A type of proof in which the conclusion is shown to be true indirectly from statements of the proof. The statements show that all of the alternatives to the conclusion are shown to be false. The conclusion is shown to be true because the assumption that its negation is true leads to a contradiction.

Paragraph Proof – A type of proof format in which the steps are written out in complete sentences and arranged in a paragraph. It is less formal than a two-column proof, but identical in content.

Two-Column Proof – A type of proof format in which statements are listed in one column and the corresponding reasons for each statement being of true are listed in a second column. It is more formal than a paragraph proof, but identical in content.

13-4 Postulates and Theorems

A part of the process in creating geometric proofs requires common postulates and theorems to be defined. A **definition** is a statement of precise meaning. A **postulate** is a statement that is accepted as true without proof. An **axiom**, or **common notation**, is a self-evident assumption. A **theorem** is statement that can be deduced or proven from definitions, postulates, axioms, and other previously proven theorems. In addition to the geometric definitions, postulates, and axioms of Euclid and Hilbert shown in the Geometric Constructions chapter, the following table of algebraic postulates are also important for geometric proofs.

Postulate	Description
Addition	If $A = B$, then $A + C = B + C$.
Subtraction	If $A = B$, then $A - C = B - C$.
Multiplication	If $A = B$, then $A \cdot C = B \cdot C$.
Division	If $A = B$ and $C \neq 0$, then $A / C = B / C$.
Halves	If $A = B$, then $A / 2 = B / 2$.
Doubles	If $A = B$, then $2 \cdot A = 2 \cdot B$.
Substitution	If $A = 3$ and $A + B = 7$, then $3 + B = 7$.
Reflexive	$A = A$.
Symmetric	If $A = B$, then $B = A$.
Transitivity	If $A = B$ and $B = C$, then $A = C$. If $A < B$ and $B < C$, then $A < C$.
Commutative	$A + B = B + A$. $A \cdot B = B \cdot A$.
Associative	$(A + B) + C = A + (B + C)$. $(A \cdot B) \cdot C = A \cdot (B \cdot C)$.
Distributive	$A \cdot (B + C) = A \cdot B + A \cdot C$.
Identities	$A + 0 = A$. $0 + A = A$. $A \cdot 1 = A$. $1 \cdot A = A$.
Inverses	$A + (-A) = 0$. $A \cdot 1/A = 1$.

Table 13-1: Algebraic Postulates

13-5 How to Write a Proof

A proof is a sequence of logical deductions based on accepted assumptions and previously proven statements that verify a statement is true. Writing a proof involves a step-by-step process to prove a mathematical statement. There are five steps in the process as follows:

1. List the statement of the theorem – This is the mathematical statement that needs to be validated by the proof.
2. State the given information – At the start of the proof, list the facts provided by the problem.
3. Create a drawing representing the given information – Make an accurate drawing and label all points.
4. State the theorem to be proven – At the end of the proof, list the statement of the theorem.
5. Create the proof – In between the start and the end of the proof, create the statements that connect the given statements to the proven statement.

A proof can be written in the two-column format or the paragraph format. Both formats will result in identical content, but the two-column format is the most commonly used. The text of the paragraph format can become very long for a complex proof. The proof below is made using the two-column format and the paragraph format for comparison.

Proof 1 – Line Segments

Given: AC ≅ BD   Prove: AB ≅ CD

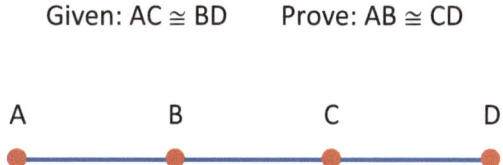

Figure 13-1: Proof 1 – Given, Prove, and Drawing

Statements	Reasons
1. AC ≅ BD	1. Given
2. AC = BD	2. Definition of congruent segments
3. AB + BC = AC; BC + CD = BD	3. Segment addition postulate
4. AB + BC = BC + CD	4. Substitution property
5. BC = BC	5. Reflexive property
6. AB = CD	6. Subtraction property
7. AB ≅ CD	7. Definition of congruent segments

Table 13-2: Proof 1 – Two-column Format

> This proof will show line segment AB is congruent to line segment CD, given the fact that line segment AC is congruent to line segment BD. The accompanying drawing depicts a line segment with end points A and D. Points B and C are located on line segment AD in between points A and D, forming line segments AB, BC, and CD. The first step of the proof is line segment AC is congruent to line segment BD. This is the given fact. The second step is AC is equal to BD. This is the definition of congruent segments. The third step is the length of line segment AB added to the length of line segment BC is equal to the length of line segment AC. Also, the length of line segment BC added to the length of line segment CD is equal to the length of line segment BD. These two statements use the segment addition postulate. The fourth step is the length of line segment AB added to the length of line segment BC is equal to the length of line segment BC added to the length of line segment CD. This uses the substitution property. The fifth step is the length of line segment BC is equal to the length of line segment BC. This is the reflexive property. The sixth step is the length of line segment AB is equal to the length of line segment CD. This is the subtraction property. The seventh step is line segment AB is congruent to line segment CD. This is the definition of congruent segments. The proof is complete.

Table 13-3: Proof 1 – Paragraph Format

Some traditional proofs, such as those in Euclid's *Elements*, will often end with the letters Q.E.D. or Q.E.F. or another short phrase. They are meant to signify a conclusion has been reached and the proof is complete. Q.E.D. stands for the Latin *quod erat demonstrandum*, meaning "that which was to have been demonstrated." It is translated from the Greek όπερ έδει δείξαι, meaning "precisely what was required to be proved." Q.E.F. stands for the Latin *quod erat faciendum*, meaning "that which was to have been done." It is translated from the Greek όπερ έδει ποιησαι, meaning "precisely what was required to be done." The phrase "Therefore, etc." is used to stand for the exact restatement of the original proposition as the concluding statement of the proof. While these abbreviations and phrases are not need when writing a proof, they are commonly used in both traditional and modern proofs.

• 13-6 Types of Proofs

There are two types of proofs, direct proofs and indirect proofs. A direct proof shows a deductive argument to be true based on the statements. An indirect proof shows the contradiction of the deductive argument to be not true. An indirect proof is more difficult to construct than a deductive proof and is less common. Examples of direct and indirect proof are shown below.

Proof 2 – Angle Bisector (direct proof)

Given: Ray AC bisects angle BAD   Prove: Angle BAC = angle DAE

Given: Ray AD bisects angle CAE

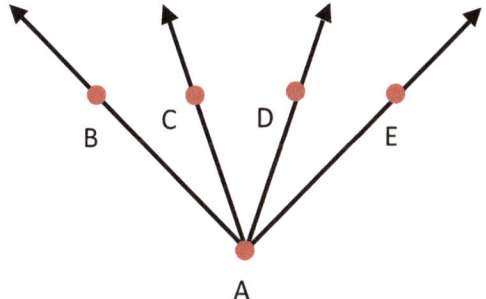

Figure 13-2: Proof 2 – Given, Prove, and Drawing

Table 13-4: Proof 2 – Angle Bisector

Statements	Reasons
1. Ray AC bisects angle BAD	1. Given
2. Ray AD bisects angle CAE	2. Given
3. BAC ≅ CAD; CAD ≅ DAE	3. Definition of angle bisector
4. BAC = CAD; CAD = DAE	4. Definition of congruent angles
5. Angle BAC = angle DAE	5. Transitive property

Proof 3 – Congruent Angles (direct proof)

Given: Angle A ≅ angle D   Prove: AE ≅ DE

Given: Angle EBC ≅ ECB

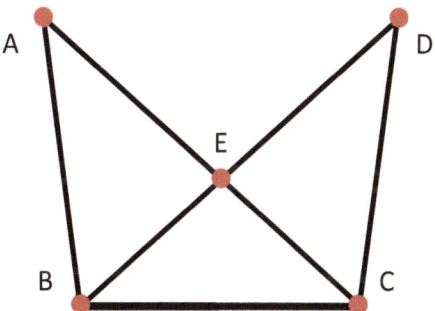

Figure 13-3: Proof 3 – Given, Prove, and Drawing

43

Table 13-5: Proof 3 – Congruent Angles

Statements	Reasons
1. Angle A ≅ angle D	1. Given
2. Angle EBC ≅ angle ECB	2. Given
3. BE ≅ CE	3. Sides opposite congruent angles are congruent
4. AEB ≅ DEC	4. Vertical angles are congruent
5. Triangle ABE ≅ triangle DCE	5. Angle-Angle-Side (AAS) congruency
6. AE ≅ DE	6. Corresponding parts of ≅ triangles are ≅ (CPCTC)

Proof 4 – Isosceles Triangles (direct proof)

Given: Segment BD is a median of triangle ABC

Given: Segment DE is a median of triangle ABD

Given: Segment DE is perpendicular to segment AB

Prove: Triangle BCD is isosceles

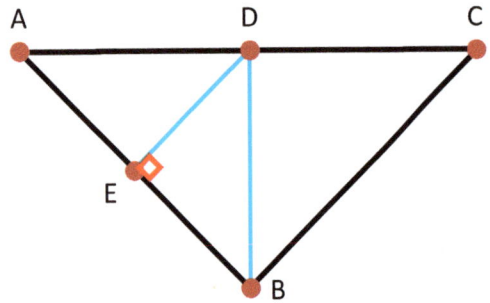

Figure 13-4: Proof 4 – Given, Prove, and Drawing

Table 13-6: Proof 4 – Isosceles Triangles

Statements	Reasons
1. Segment BD is a median of triangle ABC	1. Given
2. AD ≅ DC	2. Median bisects a side into two congruent segments
3. Segment DE is a median of triangle ABD	3. Given
4. AE ≅ EB	4. Median bisects a side into two congruent segments
5. Segment DE is perpendicular to segment AB	5. Given
6. AED is a right angle	6. Perpendicular segments form a right angle
7. BED is a right angle	7. Perpendicular segments form a right angle
8. AED ≅ BED	8. Right angles are congruent

| 9. DE ≅ DE | 9. Reflexive property |

10. Triangle AED ≅ triangle BED	10. Side-Angle-Side (SAS) congruency (steps 4,8,9)
11. AD ≅ BD	11. Corresponding parts of ≅ triangles are ≅ (CPCTC)
12. DC ≅ BD	12. Substitution property (steps 2,11)
13. Triangle BCD is isosceles	13. If two side are congruent, triangle is isosceles

Proof 5 – Triangle Angles (indirect proof)

Given: Triangle ABC is an isosceles triangle

Given: Angle A is the vertex angle

Prove: Angles B and C are congruent

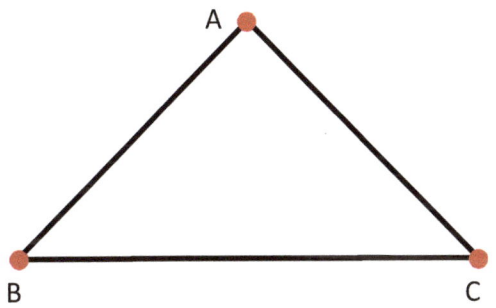

Figure 13-5: Proof 5 – Given, Prove, and Drawing

Table 13-7: Proof 5 – Triangle Angles

Statements	Reasons
1. Triangle ABC is an isosceles triangle	1. Given
2. Angle A is the vertex angle	2. Given
3. Angles B and C are not congruent	3. Indirect proof assumption
4. AB ≅ AC	4. Legs of an isosceles triangle are congruent
5. Angles B and C are congruent	5. Angles opposite congruent sides are congruent
6. Contradiction (steps 3,5)	6. Angles can not be congruent and not congruent
7. Angles B and C are congruent	7. The negation is false

Proof 6 – Triangle Bisector (indirect proof)

Given: Triangle ABC is an isosceles triangle

Given: Segment BC is the base side

Given: Segment AD is a median of triangle ABC

Prove: Segment AD is an angle bisector

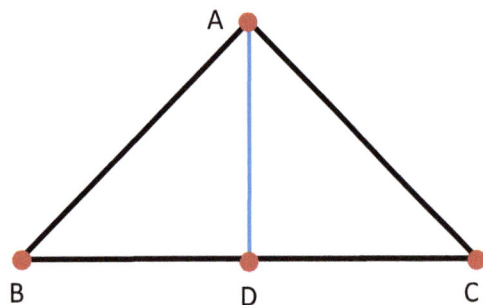

Figure 13-6: Proof 6 – Given, Prove, and Drawing

Table 13-8: Proof 6 – Triangle Bisector

Statements	Reasons
1. Triangle ABC is an isosceles triangle	1. Given
2. Segment BC is the base side	2. Given
3. AB ≅ AC	3. Legs of an isosceles triangle are congruent
4. Segment AD is not an angle bisector	4. Indirect proof assumption
5. BAD is not congruent to CAD	5. Angle A is not bisected, so angles are not congruent
6. Segment AD is a median of triangle ABC	6. Given
7. BD ≅ CD	7. Median divides a side into two congruent segments
8. Angles B ≅ angle C	8. Base angles of an isosceles triangle are congruent
9. Triangle ABD ≅ triangle ACD	9. Side-Angle-Side (SAS) congruency (steps 3,7,8)
10. Angle BAD ≅ CAD	10. Corresponding parts of ≅ triangles are ≅ (CPCTC)
11. Contradiction (steps 5,10)	11. Angles can not be congruent and not congruent
12. Segment AD is an angle bisector	12. The negation is false

Proof 7 – Triangle Incircle (indirect proof)

Given: Circle F is inscribed in triangle ABC   Prove: Angles DCF and ECF are congruent

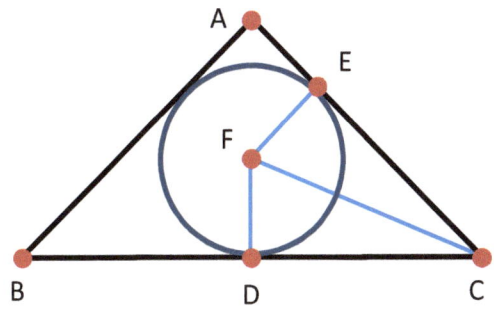

Figure 13-7: Proof 7 – Given, Prove, and Drawing

Table 13-9: Proof 7 – Triangle Incircle

Statements	Reasons
1. Circle F is inscribed in triangle ABC	1. Given
2. Angle DCF is not congruent to angle ECF	2. Indirect proof assumption
3. FD ≅ FE	3. Radii of the same circle are congruent
4. CF ≅ CF	4. Reflexive property
5. FDC and FEC are right angles	5. Radii at points of tangency form right angles
6. FDC ≅ FEC	6. Right angles are congruent
7. Triangles FDC and FEC are right triangles	7. Triangles with right angles are right triangles
8. Triangle FDC ≅ triangle FEC	8. Hypotenuse-Leg or SAS congruency (steps 3,4,6)
9. DCF ≅ ECF	9. Corresponding parts of ≅ triangles are ≅ (CPCTC)
10. Contradiction (steps 2,9)	10. Angles can not be congruent and not congruent
11. Angles DCF and ECF are congruent	11. The negation is false

13-7 Summary

A geometric proof is a series of statements based on definitions, postulates, axioms, and theorems used to show that a mathematical statement is valid or true. Two forms of deductive reasoning are the use of detachment and syllogism. A conditional statement contains a hypothesis and a conclusion. Detachment uses one conditional statement and a statement to reach a conclusion. Syllogism uses two conditional statements and reaches a conclusion by combining the hypothesis of one statement with the conclusion of another statement.

Geometric proofs are often considered the ultimate goal of learning geometry. Proofs formally present the results of mathematical thought using logic and deductive reasoning as a method for establishing the validity of ideas. Reasoning and proof offer students effective ways of communicating mathematical ideas and insights. These components, especially the ability to reason, are integral parts of understanding mathematics.

Learning geometric proofs will develop analytical thinking skills. Constructing geometric proofs requires complex organization skills to manage given information, create drawings, and to ensure the progress throughout the task. Proofs require logical thinking and a higher level of thinking to gather and use all of the relevant geometric definitions, postulates, axioms, and theorems in a valid deductive argument that proves a mathematical statement.

A part of the process in creating geometric proofs requires common postulates and theorems to be defined. A proof is a sequence of logical deductions based on accepted assumptions and previously proven statements that verify statement is true. Writing a proof involves a step-by-step process to prove a mathematical statement.

A proof can be written in the two-column format or the paragraph format. Both formats will result in identical content, but the two-column format is the most commonly used. There are two types of proofs, direct proofs and direct proofs. A direct proof shows a deductive argument to be true based on the statements. An indirect proof shows the contradiction of the deductive argument to be not true.

CHAPTER 13 — Chapter Test

Grading Scale: One point for each correct answer.

Excellent = 13-14, Good = 12, Average = 10-11, Fair = 9, Poor = 0-8

13-1 Introduction

Match definitions and terms.

A = Geometric Proof   B = Conjecture   C = Theorem   D = Argument
E = Conditional Statement   F = Detachment   G = Syllogism

1. An unproven proposition or hypothesis that is believed to be true. ____
2. A collection of reasons or evidence used to prove a conjecture. ____
3. Statement that contains a hypothesis and a conclusion. ____
4. A series of statements to show that a mathematical statement is valid or true. ____
5. Two conditional statements, combines hypothesis of one and conclusion of other. ____
6. Uses one conditional statement and a statement to reach a conclusion. ____
7. A conjecture that proven to be true. ____

13-3 Terms

Match definitions and terms.

A = Auxiliary Lines   B = Contradiction   C = Direct Proof   D = Geometric Proof
E = Indirect Proof   F = Paragraph Proof   G = Two-Column Proof

1. Steps are written out in complete sentences and arranged in a paragraph. ____
2. The conclusion is shown to be true directly from statements of the proof. ____
3. The conclusion is shown to be true indirectly from statements of the proof. ____
4. Lines that are added to a given drawing which help to demonstrate something. ____
5. Statements are in one column and corresponding reasons in a second column. ____
6. A statement that is always false. ____
7. A series of statements to show that a mathematical statement is valid or true. ____

CHAPTER 14 — Assessment

Final Exam: Chapter 1 – Chapter 13

Grading Scale: Questions 1-153 are 1 point each. Questions 154-156 are 5 points each.

Excellent = 152-168, Good = 135-151, Average = 118-134, Fair = 101-117, Poor = 0-100

Chapter 1 – Concepts and Standards			
Number	Question	Choice 1	Choice 2	Choice 3
1.	Type of geometry where the shortest distance is a curved line.	A. Plane	B. Solid	C. Spherical
2.	Type of geometry with x-axis, y-axis, and z-axis.	A. Plane	B. Solid	C. Spherical
3.	Type of geometry which can only exist on a flat surface.	A. Plane	B. Solid	C. Spherical
4.	The common endpoint of the rays or lines of an angle.	A. Line segment	B. Vertex	C. Ray
5.	Object with zero dimensions.	A. Polygon	B. Line	C. Point
6.	Object with three dimensions.	A. Prism	B. Tetragon	C. Line
7.	Not one of the five object identification steps.	A. Two or three dimensions	B. Calculation	C. Comparison
8.	An example of a polytope.	A. Circle	B. Cone	C. Square
9.	A quadrilateral with two pairs of adjacent equal sides.	A. Kite	B. Rhombus	C. Parallelogram
10.	Not one of the five levels of geometric understanding.	A. Visualization	B. Deduction	C. Association
11.	Not one of the five learning phases.	A. Explicitation	B. Recognition	C. Integration
12.	Not one of the three main phases of the Connected Mathematics Project (CPM).	A. Learn	B. Explore	C. Summarize

Table 14-1: Chapter 1 – Concepts and Standards

Chapter 2 - Angles			
Number	Question	Choice 1	Choice 2	Choice 3
13.	Two perpendicular lines that divide a plane into quadrants.	A. Quadrant axis	B. Coordinate axis	C. Terminal axis
14.	Objects which have the same shape and size.	A. Equal	B. Similar	C. Congruent
15.	Isometry transformation used to slide an object.	A. Rotation	B. Reflection	C. Translation
16.	The measure of an angle created by an arc that has the same length as the circle's radius.	A. Radian	B. Degree	C. Turn
17.	Angle equal to π radians.	A. Straight	B. Obtuse	C. Full
18.	Angle less than 90 degrees.	A. Right	B. Acute	C. Zero
19.	Angle greater than 1/2 turn and less than 1 turn.	A. Obtuse	B. Straight	C. Reflex
20.	Two angles that sum to a full angle.	A. Complementary	B. Supplementary	C. Explementary
21.	Lines that are always the same distance apart and do not intersect.	A. Parallel	B. Perpendicular	C. Transversal
22.	Angles formed between the two lines that are crossed by the transversal line.	A. Interior	B. Exterior	C. Vertical
23.	The sum of four angle created by intersecting lines.	A. Right angle	B. Straight angle	C. Full angle
24.	Two lines crossed by a transversal are not parallel.	A. Any pair of alternate interior angles is equal.	B. Only the vertical angles are equal.	C. Any pair of corresponding angles is equal.

Table 14-2: Chapter 2 – Angles

Number	Question	Choice 1	Choice 2	Choice 3
	Chapter 3 - Polytopes			
25.	Multi-dimensional solid with flat sides.	A. Hyperplane	B. Polytope	C. Half-space
26.	Two-dimensional space hyperplane.	A. Point	B. Line	C. Plane
27.	Triangle is the simplest form of this shape.	A. Polychoron	B. Polyhedron	C. Polygon
28.	A point is this type of polytope.	A. Monad	B. Polytelon	C. Polygon
29.	The simplest possible shape for a dimension.	A. Simplex	B. Orthoplex	C. Hypercube
30.	The sides are in parallel pairs orthogonal or perpendicular to each other.	A. Simplex	B. Orthoplex	C. Hypercube
31.	The vertices are in pairs orthogonal or perpendicular to each other.	A. Simplex	B. Orthoplex	C. Hypercube
32.	An apeirotope or an infinite sided n-polytope	A. Polyhedron	B. Polyxennon	C. Apeirohedron
33.	A curve that does not self-intersect.	A. Simple	B. Topological	C. Polygonal

Table 14-3: Chapter 3 – Polytopes

Chapter 4 - Polygons

Number	Question	Choice 1	Choice 2	Choice 3
34.	Polygon shape with no internal angles greater than 180 degrees.	A. Concave	B. Convex	C. Complex
35.	Polygon with all of the vertices lying on a single circle.	A. Non-cyclic	B. Star	C. Cyclic
36.	Polygons that is equilateral and equiangular.	A. Regular	B. Irregular	C. Complex
37.	A line connecting two vertices that are not a side.	A. Internal angle	B. Exterior angle	C. Diagonal
38.	The sum of the interior angles of a 4 sided polygon.	A. 360 degrees	B. 540 degrees	C. 720 degrees
39.	The number of triangles in a 5 sided polygon.	A. 2	B. 3	C. 4
40.	The common or preferred name for a 6 sided polygon.	A. Hexangle	B. Hexalateral	C. Hexagon
41.	The distance from the center of a polygon to the midpoint of a side.	A. Apothem	B. Radius	C. Perimeter
42.	Circle outside of a polygon that intersects each of the vertices.	A. Incircle	B. Circumcircle	C. Vertex Circle
43.	In a triangle, a line from a vertex to midpoint of the opposite side that divides the base into two equal parts.	A. Median	B. Bisector	C. Altitude
44.	In a triangle, a line divides a triangle into two opposite facing right triangles.	A. Median	B. Bisector	C. Altitude
45.	In a triangle, a transitive line that divides the angle into two equal parts.	A. Median	B. Bisector	C. Altitude

Table 14-4: Chapter 4 – Polygons

Chapter 5 – Triangles and Quadrilaterals

Number	Question	Choice 1	Choice 2	Choice 3
46.	Triangle that has all three sides of unequal length.	A. Isosceles	B. Scalene	C. Equilateral
47.	Triangle that has one angle equal to 90 degrees.	A. Right	B. Obtuse	C. Oblique
48.	Hypotenuse length in a triangle with legs of 6 and 8 units.	A. 12 units	B. 10 units	C. 8 units
49.	Triangles that have the exact same shape, but may not be the same size.	A. Congruent	B. Similar	C. Oblique
50.	Triangles that have the exact same shape and size.	A. Similar	B. Isosceles	C. Congruent
51.	Equality condition not sufficient to prove congruency for a pair of triangles.	A. (AAS) or angle-angle-side	B. (SAS) or side-angle-side	C. (AAA) or angle-angle-angle
52.	The three perpendicular bisectors meet in a single point.	A. Orthocenter	B. Circumcenter	C. Incenter
53.	The three altitudes intersect in a single point.	A. Orthocenter	B. Circumcenter	C. Incenter
54.	The three angle bisectors intersect in a single point.	A. Orthocenter	B. Circumcenter	C. Incenter
55.	The three medians intersect in a single point.	A. Centroid	B. Morley center	C. Fermat center
56.	A point of a triangle that is a minimum distance from the three vertices.	A. Centroid	B. Morley center	C. Fermat center
57.	Quadrilateral with one pair of parallel sides and one pair of equal sides.	A. Isosceles trapezoid	B. Rhombus	C. Rectangle
58.	Quadrilateral without any special properties, such as parallel sides, equal sides, or congruent angles	A. Kite	B. Scalene	C. Trapezoid
59.	A quadrilateral that is both cyclic and tangential.	A. Bicentric	B. Orthodiagonal	C. Inscriptible
60.	The area of a trapezoid with bases 3 and 5 units, and height 4 units.	A. 12 square units	B. 20 square units	C. 16 square units.

Table 14-5: Chapter 5 – Triangles and Quadrilaterals

	Chapter 6 - Polyhedron			
Number	Question	Choice 1	Choice 2	Choice 3
61.	Polyhedron with edges only, without surface area or volume.	A. Solid shape	B. Hollow shape	C. Wire frame
62.	Polyhedron with edges and surface area, without volume.	A. Solid shape	B. Hollow shape	C. Wire frame
63.	Polyhedron with edges, surface area, and volume.	A. Solid shape	B. Hollow shape	C. Wire frame
64.	Polyhedron with a surface that is not uniformly flat or is like a star.	A. Concave	B. Convex	C. Complex
65.	Polyhedron with ten sides.	A. Octahedron	B. Icosahedron	C. Decahedron
66.	Hexahedron with five vertices.	A. Hemiobelisk	B. Tetragonal antiwedge	C. Triangular dipyramid
67.	A polyhedron whose faces are all equilateral triangle.	A. Trihedron	B. Deltahedron	C. Tetrahedron
68.	Polyhedron where there is only one type of edge to an object, such as a hexagon face meeting another hexagon face.	A. Isohedral	B. Isogonal	C. Isotoxal
69.	Polyhedron where all of the faces must be congruent.	A. Isohedral	B. Isogonal	C. Isotoxal
70.	Polyhedron where each vertex is surrounded by the same kinds of face in the same or reverse order and with the same angles between the corresponding faces.	A. Isohedral	B. Isogonal	C. Isotoxal
71.	A category of polyhedral with isogonal, isotoxal, and isohedral symmetry.	A. Regular	B. Noble	C. Quasi-regular
72.	The dual polyhedron of a cube.	A. Octahedron	B. Icosahedron	C. Tetrahedron

Table 14-6: Chapter 6 – Polyhedron

Chapter 7 – Polyhedron Solids – Part 1

Number	Question	Choice 1	Choice 2	Choice 3
73.	Platonic solid made up of eight equilateral triangles.	A. Tetrahedron	B. Octahedron	C. Icosahedron
74.	The angle between the segments joining the center and the vertices.	A. Dihedral	B. Vertihedral	C. Polyhedral
75.	The process of extending edges or faces until they meet to form a new polygon or polyhedron.	A. Stellation	B. Extension	C. Concavation
76.	Kepler-Poinsot solid with twenty vertices.	A. Small Stellated Dodecahedron	B. Great Stellated Dodecahedron	C. Great Dodecahedron
77.	Archimedean solid process that is a form of expansion in which all of the faces are slightly rotated in the same direction. The empty spaces are filled with triangles.	A. Snubification	B. Rectification	C. Truncation
78.	Archimedean solid that does not have five edges meeting at each vertex.	A. Snub cuboctahedron	B. Truncated tetrahedron	C. Snub dodecahedron
79.	Catalan solid with 120 faces.	A. Deltoidal icositetrahedron	B. Pentakis dodecahedron	C. Disdyakis triacontahedron
80.	Catalan solid made up of right isosceles triangles.	A. Tetrakis hexahedron	B. Triakis octahedron	C. Disdyakis dodecahedron
81.	Johnson solid descriptive term or adjective for a prism that is joined to the base of a solid or between the bases of a solid.	A. Augmented	B. Sphenoid	C. Elongated
82.	Polyhedron solids which are non-convex regular polyhedra	A. Kepler-Poinsot solids	B. Archimedean solids	C. Catalan solids
83.	Polyhedron solids which are dual polyhedrons to the Archimedean solids.	A. Johnson solids	B. Catalan solids	C. Kepler-Poinsot solids
84.	Polyhedron solids which are convex regular polyhedra.	A. Archimedean solids	B. Johnson solids	C. Platonic solids

Table 14-7: Chapter 7 – Polyhedron Solids – Part 1

Number	Question	Choice 1	Choice 2	Choice 3
85.	Pyramid type that is not regular pyramid.	A. Square pyramid	B. Pentagonal pyramid	C. Hexagonal pyramid
86.	The volume for a square base pyramid, with base length of 6 units and the height of 4 units.	A. 72 cubic units	B. 48 cubic units	C. 36 cubic units
87.	The volume for a square base bipyramid, with base length of 8 units and the height of 6 units.	A. 256 cubic units	B. 192 cubic units	C. 128 cubic units
88.	Shape that composes the faces of a trapazohedron.	A. Trapezoid	B. Kite	C. Parallelogram
89.	The surface area of a right square pyramidal frustum, with bottom base length of 4 units, top base length of 1 unit, and slant height of 3 units.	A. 18 square units	B. 20 square units	C. 30 square units
90.	Two frustra joined together.	A. Bifrustum	B. Antipryramid	C. Antiprism
91.	A regular prism can not have this shape as a lateral side.	A. Rectangle	B. Trapezoid	C. Parallelogram
92.	The volume of a rectangular cubic prism, with a base length of 2 unit, base width of 4, and height of 6 units.	A. 24 cubic units	B. 36 cubic units	C. 48 cubic units
93.	Shape that composes the faces of an antiprism.	A. Rectangle	B. Kite	C. Triangle
94.	A wedge can not have this shape as a side.	A. Square	B. Triangle	C. Trapezoid
95.	A polyhedron that is derived by tunneling into a convex polyhedron to remove a section.	A. Quasi-convex	B. Semi-regular	C. Anti-convex
96.	Polyhedron composed of a number of interpenetrating polyhedra, either of the same of several different types, which share a common center.	A. Toroidal	B. Compound	C. Augmented

Table 14-8: Chapter 8 – Polyhedron Solids – Part 2

Chapter 9 – Two Dimensional Non-polytopes

Number	Question	Choice 1	Choice 2	Choice 3
97.	The boundary of a double cone.	A. Conical surface	B. Conic section	C. Conic edge
98.	Conic section that is not a closed curve.	A. Circle	B. Ellipse	C. Parabola
99.	A line that intersects a circle in two different points.	A. Tangent	B. Secant	C. Chord
100.	Part of a circle bounded by two radii and their intercepted arc.	A. Circular ring	B. Circular sector	C. Circular section
101.	Part of a circle bounded by a chord and its associated arc.	A. Circular ring	B. Circular sector	C. Circular section
102.	Ring shaped object and is the region lying between two concentric coplanar circles.	A. Circular ring	B. Circular sector	C. Circular section
103.	Annulus sector	A. Circular sector	B. Circular ring	C. Circular ring sector
104.	A line outside of an ellipse that is parallel to either of the minor axis.	A. Focus	B. Directrix	C. Semi-minor axis
105.	A point on the parabola halfway between the directrix and the focus.	A. Vertex	B. Axis point	C. Focus point
106.	The boundary of the hyperbola curve.	A. Asymptote lines	B. Transverse axis	C. Conjugate axis
107.	A shape consisting of a rectangle with semicircles attached to two opposite ends.	A. Ellipse	B. Oval	C. Stadium
108.	A shape consisting of a closed curve that may or may not be symmetrical. It does not have a precise mathematical definition.	A. Ellipse	B. Oval	C. Stadium

Table 14-9: Chapter 9 – Two Dimensional Non-polytopes

Chapter 10 – Three Dimensional Non-polytopes

Number	Question	Choice 1	Choice 2	Choice 3
109.	The shortest path between two non-antipodial points on the surface of a sphere.	A. Orthodome	B. Geodesic	C. Chord
110.	The intersection of a plane and a sphere, with the plane passing through the center of the sphere.	A. Small circle	B. Radius	C. Great circle
111.	The region of a sphere which lies above, or is cut off by, a plane	A. Spherical segment	B. Spherical sector	C. Spherical cap
112.	The region of a sphere bounded by two radii and their intercepted angle	A. Spherical segment	B. Spherical sector	C. Spherical cap
113.	The region of a sphere which lies between two parallel planes.	A. Spherical segment	B. Spherical sector	C. Spherical cap
114.	The surface of a spherical wedge.	A. Spherical frustum	B. Spherical lune	C. Spherical surface
115.	Cone in which an axis through the vertex and the center of the base, the altitude, intersects the base perpendicularly.	A. Oblique cone	B. Center cone	C. Right cone
116.	The volume of a cylinder compared to the volume of a cone with the same base and height.	A. Equal volume	B. Three times larger	C. Half as large
117.	Constructed by cutting off the top of a cone with a plane parallel to the base or by reducing the diameter of one base of a cylinder.	A. Conical frustum	B. Capsule	C. Torus
118.	Formed by revolving a circle in three-dimensional space around the z-axis.	A. Conical frustum	B. Capsule	C. Torus
119.	A three-dimensional stadium or a cylinder with two hemispherical caps.	A. Conical frustum	B. Capsule	C. Torus
120.	The area enclosed by a sphere.	A. Ball	B. Disk	C. Spherical Center

Table 14-10: Chapter 10 – Three Dimensional Non-polytopes

Chapter 11 – Spherical Geometry

Number	Question	Choice 1	Choice 2	Choice 3
121.	These objects do exist on a sphere.	A. Similar triangles	B. Parallel lines	C. Points
122.	Part of a geodesic, with points marking the beginning and ending locations.	A. Line segment	B. Geodesic segment	C. Ray
123.	A path of constant curvature.	A. Geodesic	B. Small circle	C. Meridian
124.	The arc of the great circle between the two points.	A. Geodesic	B. Small circle	C. Meridian
125.	The central angle formed at the center of the sphere by the intersection of two great circles that form the sides at a vertex.	A. Polyhedral	B. Vertihedral	C. Spherical
126.	Spherical polygon created by the intersection of two great circles.	A. Spherical biangle	B. Spherical triangle	C. Spherical quadrilateral
127.	The amount by which the sum of the angles of a polygon on a sphere exceeds the sum of the angles of a polygon with the same number of sides in a plane.	A. Polygonal deficit	B. Spherical excess	C. Angle sum difference
128.	Each angle of a spherical polygon compared to each angle of a plane polygon with the same number of sides.	A. Larger	B. Equal	C. Smaller
129.	Regular spherical polygon that can not be used for tessellation to completely cover the surface of a sphere.	A. Triangle	B. Pentagon	C. Hexagon
130.	Congruency for spherical triangles, but not for plane triangle.	A. (SAS) or side-angle-side	B. (AAA) or angle-angle-angle	C. (SSS) or side-side-side
131.	The number of spherical triangles resulting from three mutually perpendicular planes passing through the center of a sphere, cutting the surface of the sphere.	A. Three	B. Six	C. Eight
132.	The angles of a spherical triangle are measured at this location.	A. Center of the sphere	B. Equator of the sphere	C. Pole of the sphere

Table 14-11: Chapter 11 – Spherical Geometry

Chapter 12 – Geometric Constructions

Number	Question	Choice 1	Choice 2	Choice 3
133.	Geometry tool used to measure angles.	A. Straightedge	B. Compass	C. Protractor
134.	Geometry tool used to construct straight lines and does not have any distance markings.	A. Straightedge	B. Compass	C. Protractor
135.	Geometry tool used to construct circles or arcs.	A. Straightedge	B. Compass	C. Protractor
136.	Geometric construction method that uses a straightedge and fixed compass.	A. Modified traditional	B. Poncelet-Steiner	C. Modern traditional
137.	Geometric construction method that uses a straightedge and collapsing compass.	A. Mohr-Mascheroni	B. Euclidian	C. Modern
138.	A statement of precise meaning.	A. Definition	B. Axiom	C. Postulate
139.	A statement that is accepted as true without proof.	A. Definition	B. Axiom	C. Postulate
140.	A self-evident assumption.	A. Definition	B. Axiom	C. Postulate
141.	Euclid definition: A plane figure contained by one line such that all the straight lines falling upon it from one point among those lying within the figure are equal to one another.	A. Square	B. Circle	C. Equilateral triangle
142.	Hilbert assumption group that describes collinearity and intersection.	A. Incidence axioms	B. Congruence axioms	C. Continuity axioms
143.	Hilbert definition: If a point B lies between a point A and a point C then the points A, B, C are three distinct points of a line, and B then also lies between C and A.	A. Continuity axiom	B. Parallel axiom	C. Order axiom
144.	This is not one of the five basic traditional constructions.	A. Construct a line drawn arbitrarily tangent to a circle.	B. Construct a circle with a given center and given radius.	C. Determine the points of intersection of two given circles.

Table 14-12: Chapter 12 – Geometric Constructions

Chapter 13 – Geometric Proofs

Number	Question	Choice 1	Choice 2	Choice 3
145.	An unproven proposition or hypothesis that is believed to be true.	A. Theorem	B. Conjecture	C. Conditional statement
146.	A collection of reasons or evidence used to prove a conjecture.	A. Detachment	B. Syllogism	C. Argument
147.	A statement that is always false.	A. Contradiction	B. Auxiliary	C. Negation
148.	Algebraic Postulate: If A = B and B = C, then A = C.	A. Associative	B. Transitivity	C. Commutative
149.	This is not one of the five steps in the process to prove a mathematical statement.	A. State the given information.	B. List the statement of the theorem.	C. Create postulates to support theorem.
150.	Not a type of proof.	A. Postulate	B. Indirect	C. Direct
151.	Based on figure 14-1 below, select the correct proof.	A. Choice A	B. Choice B	C. Choice C
152.	Based on figure 14-2 below, select the correct proof.	A. Choice A	B. Choice B	C. Choice C
153.	Based on figure 14-3 below, select the correct proof.	A. Choice A	B. Choice B	C. Choice C
154.	Based on figure 14-4 below, write the correct proof.	Write the correct proof. (5 points)		
155.	Based on figure 14-5 below, write the correct proof.	Write the correct proof. (5 points)		
156.	Based on figure 14-6 below, write the correct proof.	Write the correct proof. (5 points)		

Table 14-13: Chapter 13 – Geometric Proofs

Proof 1 – Perpendicular Lines

Given: Angle EAB ≅ Angle EAD   Prove: AC is perpendicular to BD

Given: Angle ECB ≅ Angle ECD

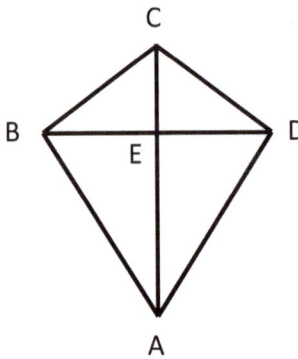

Figure 14-1: Proof 1 – Given, Prove, and Drawing

Statements	Reasons
1. Angle EAB ≅ Angle EAD	1. Given
2. Angle ECB ≅ Angle ECD	2. Given
3. AC ≅ AC	3. Reflexive property
4. Triangle ABC ≅ Triangle ADC	4. Side-Angle-Side (SAS) postulate
5. AB ≅ AD	5. CPCTC
6. AE ≅ AE	6. Reflexive property
7. Triangle ABE ≅ Triangle ADE	7. Angle-Side-Angle (ASA) postulate
8. Angle AEB ≅ Angle AED	8. CPCTC
9. AC is perpendicular to BD	9. If two lines form congruent adjacent angles, then the lines are perpendicular.

Table 14-14: Proof 1 – Perpendicular Lines – Choice A

Statements	Reasons
1. Angle EAB ≅ Angle EAD	1. Given
2. Angle ECB ≅ Angle ECD	2. Given
3. AC ≅ AC	3. Reflexive property
4. AB ≅ AD	4. CPCTC
5. AE ≅ AE	5. Reflexive property
6. Angle AEB ≅ Angle AED	6. CPCTC
7. Triangle ABE ≅ Triangle ADE	7. Side-Angle-Side (SAS) postulate
8. Triangle ABC ≅ Triangle ADC	8. Angle-Side-Angle (ASA) postulate
9. AC is perpendicular to BD	9. If two lines form congruent adjacent angles, then the lines are perpendicular.

Table 14-15: Proof 1 – Perpendicular Lines – Choice B

Table 14-16: Proof 1 – Perpendicular Lines – Choice C

Statements	Reasons
1. Angle EAB ≅ Angle EAD	1. Given
2. Angle ECB ≅ Angle ECD	2. Given
3. AC ≅ AC	3. Reflexive property
4. Triangle ABC ≅ Triangle ADC	4. Angle-Side-Angle (ASA) postulate
5. AB ≅ AD	5. CPCTC
6. AE ≅ AE	6. Reflexive property
7. Triangle ABE ≅ Triangle ADE	7. Side-Angle-Side (SAS) postulate
8. Angle AEB ≅ Angle AED	8. CPCTC
9. AC is perpendicular to BD	9. If two lines form congruent adjacent angles, then the lines are perpendicular.

Proof 2 – Midpoint of a Line

Given: Angle A ≅ Angle C   Prove: E is the midpoint of BD

Given: E is the midpoint of AC

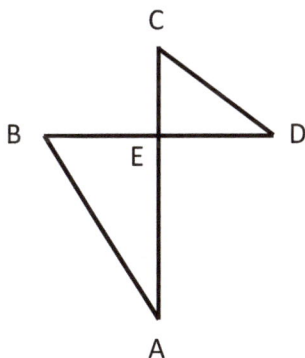

Figure 14-2: Proof 2 – Given, Prove, and Drawing

Statements	Reasons
1. Angle A ≅ Angle C	1. Given
2. E is the midpoint of AC	2. Given
3. Angle AEB ≅ Angle DEC	3. Vertical angles are congruent
4. Triangle ABE ≅ Triangle CDE	4. Angle-Side-Angle (ASA) postulate
5. BE = DE	5. CPCTC
6. E is the midpoint of BD	6. Definition of midpoint

Table 14-17: Proof 2 – Midpoint of a Line – Choice A

Table 14-18: Proof 2 – Midpoint of a Line – Choice B

Statements	Reasons
1. Angle A ≅ Angle C	1. Given
2. E is the midpoint of AC	2. Given
3. Angle AEB ≅ Angle DEC	3. Vertical angles are congruent
4. BE = DE	4. CPCTC
5. Triangle ABE ≅ Triangle CDE	5. Angle-Side-Angle (ASA) postulate
6. E is the midpoint of BD	6. Definition of midpoint

Table 14-19: Proof 2 – Midpoint of a Line – Choice C

Statements	Reasons
1. Angle A ≅ Angle C	1. Given
2. E is the midpoint of AC	2. Given
3. BE = DE	3. CPCTC
4. Triangle ABE ≅ Triangle CDE	4. Angle-Side-Angle (ASA) postulate
5. Angle AEB ≅ Angle DEC	5. Vertical angles are congruent
6. E is the midpoint of BD	6. Definition of midpoint

Proof 3 – Parallel Lines

Given: Line Segments AF, BC, and ED

Given: Angle ABC is 40 degrees

Given: Angle FED is 140 degrees

Prove: BC is parallel to ED

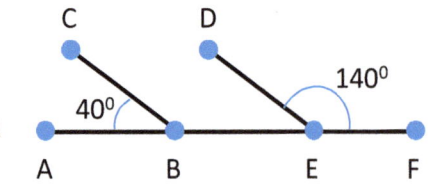

Figure 14-3: Proof 3 – Given, Prove, and Drawing

Table 14-20: Proof 3 – Parallel Lines – Choice A

Statements	Reasons
1. Line Segments AF, BC, and ED	1. Given
2. Angle ABC is 40 degrees	2. Given
3. Angle FED is 140 degrees	3. Given
4. Angle ABE is 180 degrees	4. If the sides of an angle form a straight line, then it is a straight angle and 180 degrees.
5. 140 degrees = Angle EBC	5. Subtraction property
6. 180 degrees = 40 degrees + Angle EBC	6. Substitution property
7. Angle ABE = Angle ABC + Angle EBC	7. For any angle, the measure of the whole is equal to the sum of the measures its non-overlapping parts.
8. BC is parallel to ED	8. If two lines are cut by a transversal line and corresponding angles are equal in measure, then the lines are parallel.

Table 14-21: Proof 3 – Parallel Lines – Choice B

Statements	Reasons
1. Line Segments AF, BC, and ED	1. Given
2. Angle ABC is 40 degrees	2. Given
3. Angle FED is 140 degrees	3. Given
4. Angle ABE is 180 degrees	4. If the sides of an angle form a straight line, then it is a straight angle and 180 degrees.
5. Angle ABE = Angle ABC + Angle EBC	5. For any angle, the measure of the whole is equal to the sum of the measures its non-overlapping parts.
6. 180 degrees = 40 degrees + Angle EBC	6. Substitution property
7. 140 degrees = Angle EBC	7. Subtraction property
8. BC is parallel to ED	8. If two lines are cut by a transversal line and corresponding angles are equal in measure, then the lines are parallel.

Table 14-22: Proof 3 – Parallel Lines – Choice C

Statements	Reasons
1. Line Segments AF, BC, and ED	1. Given
2. Angle ABC is 40 degrees	2. Given
3. Angle FED is 140 degrees	3. Given
4. Angle ABE is 180 degrees	4. If the sides of an angle form a straight line, then it is a straight angle and 180 degrees.
5. Angle ABE = Angle ABC + Angle EBC	5. For any angle, the measure of the whole is equal to the sum of the measures its non-overlapping parts.
6. 140 degrees = Angle EBC	6. Subtraction property
7. 180 degrees = 40 degrees + Angle EBC	7. Substitution property
8. BC is parallel to ED	8. If two lines are cut by a transversal line and corresponding angles are equal in measure, then the lines are parallel.

Proof 4 – Congruent Triangles

Given: Regular Pentagon ABCDE

Given: Line Segments AD and AE

Prove: Triangle ACD is congruent to Triangle ABE

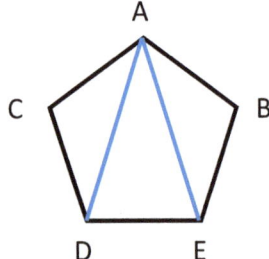

Figure 14-4: Proof 4 – Given, Prove, and Drawing

Proof 5 – Similar Triangles

Given: Regular Triangle ABC

Given: Line Segments AB and JG are parallel

Given: Line Segments BC and LI are parallel

Given: Line Segments CA and HK are parallel

Prove: Triangle ABC is similar to Triangle DEF

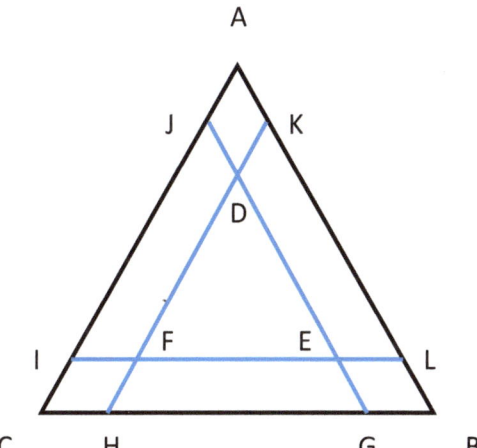

Figure 14-5: Proof 5 – Given, Prove, and Drawing

Proof 6 – Congruent Quadrilaterals

Prove: Quadrilaterals BDMJ, FHOL, and GCKN are congruent.

Given: Regular Triangles ABC, DEF, GHI, and AEI

Given: Regular Quadrilaterals BJKC, MDFL, and NOHG

Given: Regular Hexagon JMLONK

Given: All given regular polygons have sides measuring 1 unit in length.

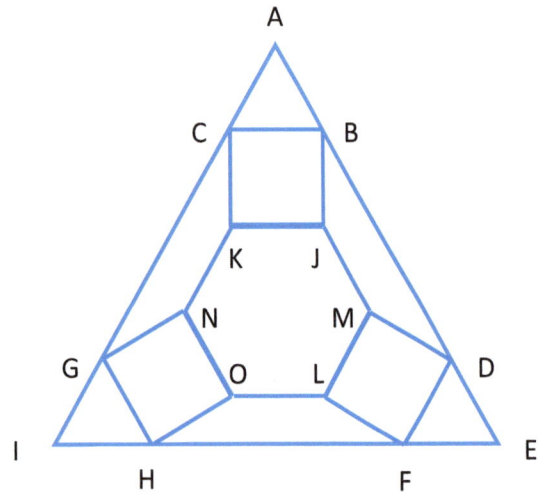

Figure 14-6: Proof 6 – Given, Prove, and Drawing

www.ingramcontent.com/pod-product-compliance
Lightning Source LLC
Chambersburg PA
CBHW051024180526
45172CB00002B/464